SpringerBriefs in Applied Sciences and Technology

Thermal Engineering and Applied Science

Series Editor
Francis A. Kulacki, University of Minnesota, USA

Pradipta Kumar Panigrahi
Krishnamurthy Muralidhar

Imaging Heat and Mass Transfer Processes

Visualization and Analysis

 Springer

Pradipta Kumar Panigrahi
Department of Mechanical Engineering
Indian Institute of Technology Kanpur
Kanpur, Uttar Pradesh, India

Krishnamurthy Muralidhar
Department of Mechanical Engineering
Indian Institute of Technology Kanpur
Kanpur, Uttar Pradesh, India

ISSN 2191-530X ISSN 2191-5318 (electronic)
ISBN 978-1-4614-4790-0 ISBN 978-1-4614-4791-7 (eBook)
DOI 10.1007/978-1-4614-4791-7
Springer New York Heidelberg Dordrecht London

Library of Congress Control Number: 2012945979

Printed on acid-free paper

Springer is part of Springer Science+Business Media (www.springer.com)

Preface

Refractive index-based optical methods are nonintrusive and inertia-free and can image cross sections of the flow field in an experimental apparatus. With the experiment suitably carried out, three-dimensional physical domains with unsteady processes can be accommodated. An introduction to the subject can be seen in companion volume entitled *Schlieren and Shadowgraph Methods in Heat and Mass Transfer*.

In this monograph, schlieren and shadowgraph imaging of transport phenomena in certain engineering applications is described. Examples relate to flow over heated bodies, convection in superposed fluid layers, convection around a crystal growing from its supersaturated solution, and jet flow patterns. In each experiment, optical imaging is shown to reveal the scalar field—temperature or concentration as a function of time. In addition, optical images reveal key transition sequences that would otherwise be hard to detect from conventional measurements. Some examples are suppression of vortex shedding from a square cylinder, stratification of the salt-saturated solution in crystal growth, thermal coupling between immiscible fluid layers, and oscillations in a negatively buoyant jet.

Optical measurement techniques, thus, emerge as a tool for whole-field imaging of complex fluid–thermal and solutal processes. They generate a vast amount of data in space and time. The data can be analyzed to retrieve relevant quantities of interest. In addition, they serve an important purpose of detecting important transport mechanisms within the domain of interest.

IIT Kanpur, India   Pradipta K. Panigrahi
                                                                    Krishnamurthy Muralidhar

Acknowledgments

We are thankful to several doctoral and master's students who worked with us on laser imaging of fluid and thermal systems. We would like to acknowledge the contributions of the following doctoral students:

1. Sunil Punjabi
2. Atul Srivastava
3. Sunil Verma
4. Surendra K. Singh
5. Anamika S. Gupta

We have drawn content from their doctoral dissertations. We have also used material from the master's theses of Atanu Phukan, Srikrishna Sahu, A.A. Kakade, Kaladhar Semwal, Rakesh Ranjan, and Vikas Kumar.

Alok Prasad, Yogendra Rathi, B.R. Vinoth, and Abhinav Parashar helped us with figures, and we thank them for their time.

Financial support from funding agencies allowed us to make the apparatus and configure the measurement systems reported in this book. We gratefully acknowledge the support received from the Department of Science and Technology, New Delhi, Board of Research in Nuclear Sciences, Mumbai, and the Ministry of Human Resource Development, New Delhi.

We thank Professor Frank Kulacki, series editor of, *SpringerBriefs in Thermal Engineering and Applied Science*, for providing us this opportunity and continuous encouragement.

We are grateful to our institute for the excellent ambience it provides and our families for continued support.

Pradipta K. Panigrahi
Krishnamurthy Muralidhar

Contents

Chapter 1
Refractive Index Methods

1.1 Introduction

Optical measurement techniques in transparent media have been discussed by the authors in a companion volume entitled *Schlieren and Shadowgraph Methods in Heat and Mass Transfer* [10]. The utility of these methods for imaging heat and mass transfer in several applications is discussed in the succeeding chapters. The optical configurations of interferometry, schlieren and shadowgraph as well as data retrieval from optical images are briefly discussed in the following section. Important references on refractive index-based methods for transport phenomena are listed at the end of the chapter.

1.2 Optical Arrangement

The layout of each of the three imaging techniques used in the present work is shown in Fig. 1.1. For measurements, a continuous wave helium-neon laser (Spectra Physics, 35 mW) serves as the coherent light source. A monochrome CCD camera (Sony, Model: XC-ST50) of spatial resolution of 768×574 pixels records the optical images of the convective field. The camera is interfaced with a personal computer (HCL, 256 MB RAM, 866 MHz) through an 8- (or 12-) bit A/D card.

The principle of operation of a Mach–Zehnder interferometer shown in Fig. 1.1a has been described earlier by Goldstein [2]. It has two mirrors and two 50% beam splitters of 150-mm diameter. The mirrors have a 99.9% silver coating and employ a silicon dioxide layer as a protective layer. The interferometer floats on pneumatic legs to isolate the optics from external vibrations. Experiments have been carried out in the infinite as well as the wedge fringe setting [9]. In the infinite fringe setting, the optical path difference between the test and the reference beams is initially zero, and bright field of constructive interference is formed. When a density disturbance is introduced in the path of the test beam, it is seen as a set of fringes over which

P.K. Panigrahi and K. Muralidhar, *Imaging Heat and Mass Transfer Processes*,
SpringerBriefs in Applied Sciences and Technology 4, DOI 10.1007/978-1-4614-4791-7_1,
© Pradipta Kumar Panigrahi and Krishnamurthy Muralidhar 2013

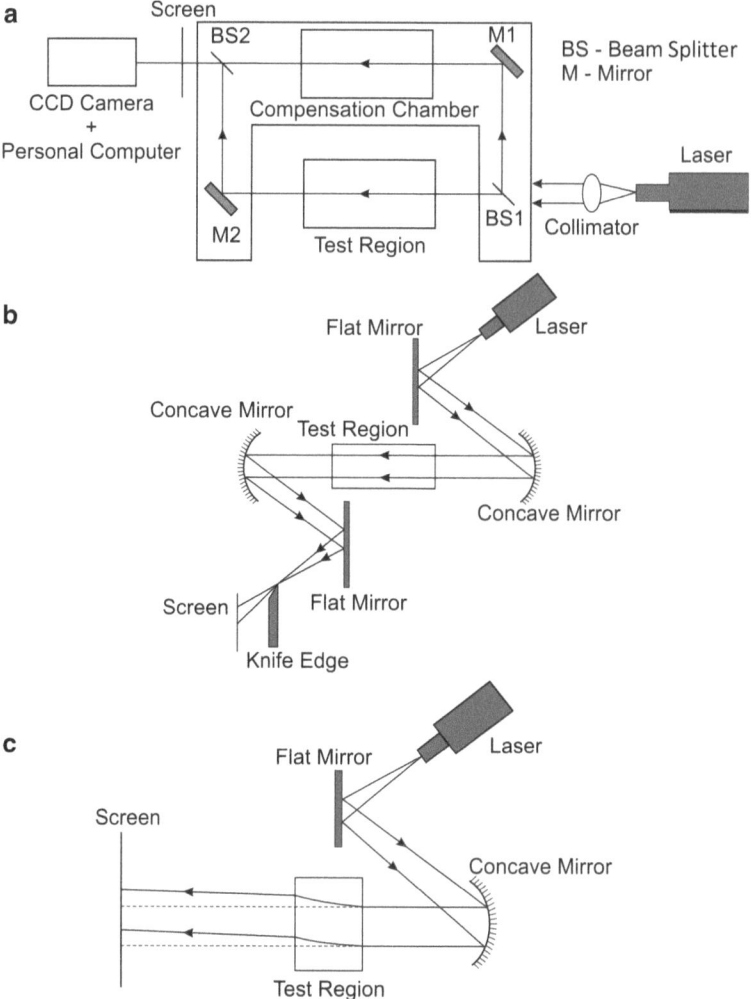

Fig. 1.1 Optical configurations of (**a**) Mach-Zehnder interferometer, (**b**) laser schlieren, and (**c**) shadowgraph

the depth-averaged density is a constant. In the wedge fringe setting, the optics is slightly misaligned to produce a set of straight fringes—horizontal or vertical. When exposed to a thermal or concentration field, the fringes are displaced to an extent depending on the change in temperature or concentration. The fringes in the wedge fringe setting of the interferometer are thus representative of changes in temperature or concentration.

The schlieren system used in the present work is of the Z-type [12, 13], as shown in Fig. 1.1b. The optics includes concave mirrors of 1.30-m focal length and 200-mm diameter. Relatively large focal lengths make the schlieren technique sensitive to the concentration gradients [10]. The knife-edge is placed at the focal length of the second concave mirror. It is positioned to cut off a part of the light focused on it, so that in the absence of any optical disturbance, the illumination on the screen is uniformly reduced. The initial intensity values in the experiment are chosen to be less than 20, on a gray scale of 0–255. The knife-edge is set perpendicular to the direction in which the density gradients are to be recorded. In the present study, the gradients are expected to be predominantly in the vertical direction, and the knife-edge has been kept horizontal.

Shadowgraph images have been recorded using the optical components employed in schlieren but without the decollimating optics and the knife-edge (Fig. 1.1c). The position of the screen on which the shadowgraph images are displayed plays an important role in data analysis [10, 12, 13]. The screen position is chosen so as to improve the image contrast, while extracting the dominant features of the flow field. In both schlieren and shadowgraph, beam refraction from regions of high concentration gradients can interfere with those passing through one of nearly constant concentration. This factor is taken into account while fixing the camera position. The initial intensity distribution in shadowgraph experiments before inserting the seed is the Gaussian variation of the laser itself, corrected for its passage through the pin-hole of the spatial filter [3].

The helium-neon laser can be replaced by a white light source and the knife-edge by a color filter in a schlieren arrangement [1, 12, 13].

1.3 Image Processing

The three optical methods generate images that are path integrals of the refractive index fields and in turn, the concentration (or temperature) distribution in the fluid medium. The integrals can be simplified if the fields being studied are strictly two-dimensional. In general, however, the local information can be extracted using principles of *tomography* [4–7]. In the present work, the images have been interpreted as carrying information of the solutal concentration (or thermal) field that is integrated along the direction of propagation of light. The analysis and interpretation of fringe patterns in interferometry has been discussed in detail by various authors [2, 9]. The approach described in [9] has been implemented in the present study.

In contrast to interferometry where information is localized at the fringes, schlieren and shadowgraph images carry information related to the local temperature or concentration in the form of an intensity distribution. The advantage here is that data is available at the pixel level of the camera. Drawbacks include the errors due to superimposed noise associated with scattering and the possibility of device saturation. In the present work, the first factor is taken to be less significant because the field variable is obtained by integrating the intensity distribution, an operation that tends to smooth noisy profiles. The second factor was circumvented by reducing the laser intensity using a neutral density filter. The local temperature and concentration was determined by numerically integrating the appropriate governing equations in cross-sectional plane of the light beam.

1.4   Data Reduction

Refractive index techniques depend on the fact that for a transparent material, refractive index (n) and density (ρ) have a unique relationship, called the Lorentz–Lorentz formula [2, 5–8, 12, 13]. It is given as

$$\frac{n^2 - 1}{\rho(n^2 + 2)} = \text{constant}. \tag{1.1}$$

In gases, $n \approx 1$ and the relationship simplifies to

$$\frac{n - 1}{\rho} = \text{constant}.$$

Hence, in gases, the derivative $dn/d\rho$ is constant. In liquids, the derivative is nearly constant if the bulk changes in density are small [11].

For moderate changes in temperature (say, up to, around $10\,^\circ\text{C}$ in air), density and temperature T are linearly related as

$$\rho = \rho_o(1 - \beta(T - T_o)), \quad \beta > 0.$$

It follows that dn/dT is constant and changes in temperature will simultaneously manifest as changes in the refractive index. This result carries over to mass transfer problems as well, where density changes occur from a solutal concentration field. In this derivation, pressure changes within the fluid region are taken to be small.

In heat transfer experiments, temperature is the only dependent variable, and density (through refractive index) relates to temperature. In mass transfer experiments such as crystal growth, temperature changes are very slow in time, and refractive index becomes a measure of concentration itself. The material property that determines the sensitivity of the optical measurement is dn/dT in heat transfer experiments and dn/dC in mass transfer. Following [2], the governing equations for each of the optical methods are summarized below; also see [10]. While the discussion is in terms of solutal concentration, it is equally applicable to imaging thermal convection.

Interferometry. Consider the passage of a laser through a test section of length L in the viewing direction. For a light source of wavelength λ, the change in concentration required per fringe shift (ΔC_ε) in the infinite fringe setting is given as

$$\Delta C_\varepsilon = \frac{\lambda/L}{dn/dC}. \tag{1.2}$$

The fringe positions are to be determined from interferogram analysis. In the wedge fringe setting, it can be shown that the fringe displacement from the initial position is proportional to the change in concentration with respect to the portion of the solution where the fringes are undisturbed. These results hold under the approximation that the light beam travels in a straight line and beam deflection and displacement effects are small.

Schlieren. Image formation in a schlieren system is due to deflection of the light beam in a variable refractive index field toward regions that have a higher refractive index. In order to recover quantitative information from a schlieren image, one has to determine the cumulative angle of refraction of the light beam emerging from the growth chamber as a function of position in the cross-sectional $x - y$ plane normal to the light beam, whose direction of propagation is along the z-coordinate. Using principles of ray optics [4], the total angular deflection δ can be expressed as [12,13]

$$\delta = \frac{1}{n_a} \int_0^L n \frac{\partial (\ln)}{\partial y} dz. \tag{1.3}$$

Here, n is the refractive index at any point in the physical domain and subscript "a" refers to the ambient. The change in the intensity field ΔI relative to the background intensity distribution I_k can now be related to the refractive index field directly as [2, 10, 12, 13]

$$\frac{\Delta I}{I_k} = \frac{f}{a_k \times n_a} \int_0^L \frac{\partial n}{\partial y} dz. \tag{1.4}$$

Here, n_a is the refractive index of the ambient in practically unity, a_k is the size of the focal spot at the knife-edge, and f is the focal length of the de collimating mirror (or lens). This equation shows that the schlieren technique records the integrated gradient of refractive index over the path of the light beam. In terms of the ray-averaged refractive index, the governing equation for the schlieren process can be derived as

$$\frac{\Delta I}{I_k} = \frac{f}{a_k} \frac{\partial n}{\partial y} L. \tag{1.5}$$

Equation (1.5) requires the approximation that changes in the light intensity occur due to beam deflection, rather than its physical displacement. The contribution of refraction of light at the confining optical windows need to be accounted for; see [10].

Shadowgraph. The shadowgraph arrangement depends on the change in the light intensity arising from beam displacement from its original path. Shadowgraph analysis requires tracing the path of individual rays through the fluid region. When subjected to linear approximations that include small displacement of the light ray, a second-order partial differential equation can be derived for the refractive index field with respect to the intensity contrast in the shadowgraph image [2, 10]. With D

as the distance of the screen from the exit plane of the experimental apparatus and Δ as the Laplace operator in the $x - y$ plane, this equation is expressed as

$$\frac{\Delta I}{I_k} = L \times D \left[\Delta \ln n(x,y) \right].$$ (1.6)

Equations (1.4) and (1.6) have to be suitably integrated to determine the refractive index and hence the concentration (or temperature) field. Integration of the Poisson equation (1.6) can be performed by a numerical technique, say the method of finite differences. When the approximations involved in the derivation of the above equations do not apply, optical techniques can be used for flow visualization alone.

References

1. K. Al-Ammar, A.K. Agrawal, S.R. Gollahalli, and D. Griffin, Application of rainbow schlieren deflectometry for concentration measurements in an axisymmetric helium jet, Experiments in Fluids, Vol. 25, pp. 89–95, 1998.
2. R.J. Goldstein, (editor), *Fluid Mechanics Measurements*, second edition, Taylor and Francis, New York, 1996.
3. J. Hecht, *The Laser Guidebook,* McGraw-Hill, New York, 1986.
4. F.A. Jenkins and H.E. White, *Fundamentals of Optics,* fourth edition, McGraw-Hill, New York, 2001.
5. M. Lehner and D. Mewes, *Applied Optical Measurements,* Springer-Verlag, Berlin, (1999).
6. W. Lauterborn and A. Vogel, Modern Optical Techniques in Fluid Mechanics, in *Annual Review Fluid Mechanics,* Vol. 16, pp 223–244, 1984.
7. F. Mayinger, Image-Forming Optical Techniques in Heat Transfer: revival by Computer-Aided Data Processing, J. Heat Transfer Trans. ASME, Vol. 115, pp 824–834, 1993.
8. F. Mayinger (Editor), *Optical Measurements: Techniques and Applications*, Springer-Verlag, Berlin, 1994.
9. K. Muralidhar, Temperature field measurement in buoyancy-driven flows using interferometric tomography, *Annual Review of Heat Transfer,* Vol. 12, pp. 265–376, 2001.
10. P.K. Panigrahi and K. Muralidhar, *Schlieren and Shadowgraph Methods in Heat and Mass Transfer,* Springer Briefs in Thermal Sciences, 2012.
11. P. Schiebener, J. Straub, J.M.H. Levelt Sengers, and J.S. Gallagher, Refractive index of water and steam as function of wavelength, temperature, and density, J. Phys. Chem. Ref. Data, Vol. 19(3), 1990.
12. G. S. Settles, *Schlieren and Shadowgraph Techniques*, Springer, Berlin, 2001, 376 pages.
13. C. Tropea, A.L. Yarin, and J.F. Foss (editors), *Springer Handbook of Experimental Fluid Mechanics,* Springer-Verlag Berlin Heidelberg (2007).

Chapter 2
Flow Past Heated Bluff Bodies

2.1 Introduction

Wakes behind heated cylinders, circular and square, have been experimentally
investigated at low Reynolds numbers. The electrically heated cylinder is mounted
in a vertical airflow facility such that the buoyancy aids the inertia of the main
flow. The dimensionless parameters, namely, Reynolds number and Richardson
number, are varied to examine flow behavior over a range of experimental conditions
from the forced to the mixed convection regime. Laser schlieren has been used
for visualization and analysis of the flow structures. The complete vortex-shedding
sequence has been recorded using a high-speed camera. The dynamical characteristics of the vortical structures—their size, shape and phase, Strouhal number, and
power spectra—are reported. On heating, the changes in the organized structures
with respect to shape, size, and their movement are readily perceived from the
instantaneous schlieren images before they reduce to a steady plume. The effect
of cylinder orientation and oscillation on the wake are also discussed.

2.2 Physical Mechanisms

Flow over bluff bodies has received immense attention from numerical and experimental fluid dynamicists [1, 8–10, 18, 19]. Examples of flow past heated circular
and square cylinders are seen in heat exchangers, cooling of electronic components,
and chemical reactors. The wake of a cylinder is unstable with increasing Reynolds
numbers and instability results in the formation of a Karman vortex street, namely,
a regular pattern of alternately shed vortices. The instability is primarily responsible
for unsteady forces and heat transfer from a cylindrical structure. Clear understanding of wake instability mechanism is important for flow control applications.
Geometry of cylinder cross-section such as square or circular affects the wake

P.K. Panigrahi and K. Muralidhar, *Imaging Heat and Mass Transfer Processes*,
SpringerBriefs in Applied Sciences and Technology 4, DOI 10.1007/978-1-4614-4791-7_2,
© Pradipta Kumar Panigrahi and Krishnamurthy Muralidhar 2013

properties. For a square cross-section, its orientation with respect to the mean flow is also an important parameter.

Most studies of bluff body wakes have focused on an isothermal fluid in which no temperature difference is maintained between the body and the ambient. However, quite a few engineering applications require the knowledge of flow behavior in wakes of heated objects [6, 16, 21, 31]. When the object is heated with respect to the incoming flow, the following possibilities arise: forced convection regime, where density changes are small; mixed convection, where density changes are significant; and free convection, where density changes (and, to an extent, changes in viscosity) entirely determine the flow field. The influence of heating also depends on orientation of the flow direction with respect to the direction of gravity. Richardson number is an appropriate dimensionless parameter to characterize buoyancy effects at a given Reynolds number. The square cylinder is fundamentally different from that of the circular cylinder due to its fixed points of separation. Therefore, it is of interest to compare the effect of heating on vortex-shedding mechanism of the circular cylinder with that of the square. When probes such as hot-wire is used, there are inherent difficulties in conducting experiments with a strongly heated cylinder. In this context, optical techniques are non-intrusive and are better suited. The laser schlieren technique is used for characterization of the heated wake of circular and square cylinders. The instantaneous thermal field, RMS fluctuations, power spectra, convection velocity, phase shift, and time traces are presented for understanding the interaction of vortical structures in the near wake of heated cylinders. Details of flow mechanism that leads to the suppression of instability in heated wakes are reported.

Additional studies reported in this chapter include buoyancy-affected wake of an oscillating cylinder [26] and the wake of a square cylinder inclined with respect to the main flow [23].

2.3   Experimental Setup

The schematic diagram of the experimental apparatus and the optical measurement system used in the present study is shown in Fig. 2.1. The setup comprises a vertical flow facility, a horizontal heated cylinder, laser schlieren apparatus, and image acquisition system. The details of the flow facility with the instrumentation used are discussed in the present section. The flow facility has also been used for measurements in the wake of an oscillating cylinder as well as a square cylinder misaligned with respect to the main flow direction.

Experiments were performed in a vertical test cell that resembles an open circuit wind tunnel. The test cell is made of Plexiglas and consists of a settling chamber, honeycomb structure, antiturbulence wire screens, contraction cone, test region, and an outflow section. The contraction cone is connected to the settling chamber and the test section with a contraction ratio of 4:1. A 2.25-kW centrifugal blower with its motor speed regulated by frequency-based drive (VICTOR G1000, *Kirloskar Electric*) was used to maintain steady flow in the vertically upward direction.

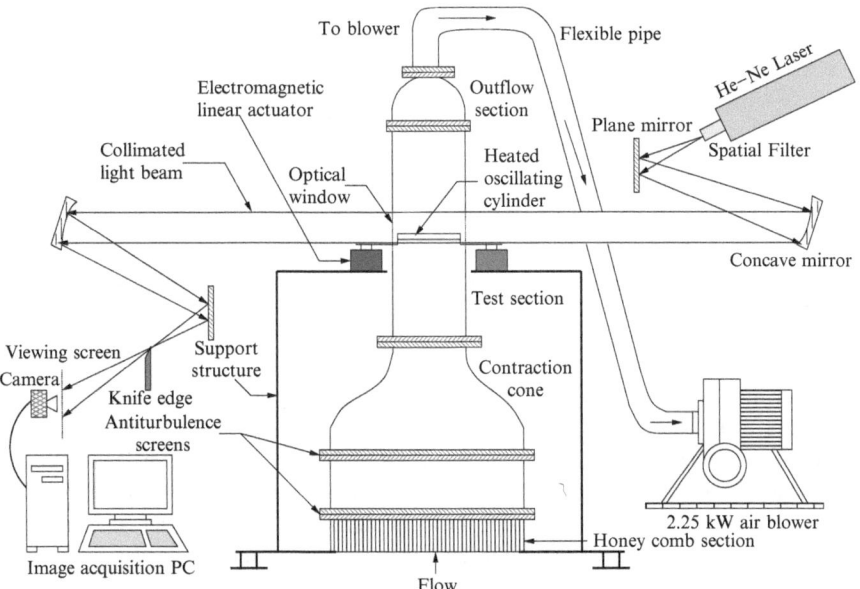

Fig. 2.1 Schematic diagram of the experimental setup with Z-type laser schlieren apparatus

Experiments were performed in the velocity range of 0.14–0.27 m/s in 0.95-m-long test section of $0.4 \times 0.4 \, \text{m}^2$ cross-section. The free stream turbulence was less than 0.4%, and flow uniformity was better than 1% over 95% of the width (outside the wall boundary layers) for the velocity range considered. The free stream velocity and velocity distribution across the wake were measured by a pitot static tube connected to a high-resolution digital micromanometer (FC012, *Furness Controls* 1.999 Pascal) with a resolution of 0.001 Pascal. A one-wire hot-wire probe (5-μm diameter, platinum-plated tungsten wire) in conjunction with a constant temperature anemometer (CTA, Model 1050, *TSI*) was used to measure turbulence intensity and power spectra. The hot-wire signal over a time interval of 20 s at 1 kS/s was acquired via a PC through a 12-bit A/D card (PCI-MIO-16E-4, *National Instruments*) with LabVIEW as the data collection software. A spectrum analyzer was also used for real-time spectral analysis of hot-wire signals to compute the power spectrum and hence the vortex-shedding frequency.

Test cylinders, one circular (6.2-mm diameter, 390 mm long) and the other square (6.7-mm edge, 390 mm long), were employed as model bluff objects. Each cylinder was placed horizontally with its major axis and one of the faces of square cylinder at right angles to the main flow direction. The cylinders spanned the full width of the test section resulting in an aspect ratio of 65 (circular) and 60 (square) with a corresponding blockage ratio of 1.55% and 1.68%, respectively. The cylinders made of copper were well polished to remove oil and dust buildup. The square cylinder was carefully machined to produce a square section with sharp edges. Both cylinders were internally heated with a Nichrome wire positioned at the axis of the

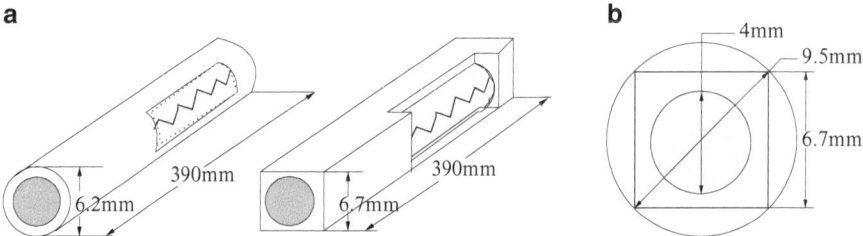

Fig. 2.2 (**a**) Dimensions of the circular and square cylinders and (**b**) fabrication of square section cylinder from circular cylinder

cylinder tube and through-out the length of the cylinder (Fig. 2.2a). The wire was insulated from the copper body by packing mica powder in the gap. The power to the heater was supplied by a regulated DC power supply (*Elnova*). To reduce end heat losses, two small Teflon plugs of 5-mm length were attached to the ends of the cylinder. Two sharp metallic holders supported the cylinder at each plug through small openings in the tunnel. The surface temperature was measured using a pre-calibrated, 36-gauge (*AWG*, 0.127 mm) chromel–alumel (type K) thermocouple flush with the cylinder using a heat-conducting epoxy. Using a DC power supply and temperature controller in a closed loop with solid state relay circuit, the temperature of the cylinder was maintained constant with an accuracy of $\pm 0.2\,^{\circ}$C. Temperature was measured at various axial and circumferential locations to check uniform heating of the cylinder surface. The temperature was uniform to within $\pm 0.1\,^{\circ}$C around the cylinder circumference and at all faces of the square cylinder. Temperature non-uniformities of up to $\pm 0.3\,^{\circ}$C were recorded along the cylinder axis for the highest temperature considered.

Figure 2.2b shows a schematic drawing to fabricate the square section of 6.7-mm edge from a thick gauge copper tube with 9.5-mm outer diameter and 4-mm internal diameter. The rounded material from the four sides was carefully removed by milling and then finished to generate a perfect square section. The origin of the wake coordinate system is fixed over the top surface of the cylinder at the midspan as shown in Fig. 2.3. The x-axis is directed upward (the streamwise direction) along the vertical. The y-axis is perpendicular (the transverse direction) to the flow and the cylinder axis. The z-axis lies on top surface of the cylinder parallel to the cylinder axis (the spanwise direction). Two linear electromagnetic actuators (SP2, *Spranktronics*) driven by a dual channel power oscillator were used to generate controlled oscillations of the cylinder. The actuators were mounted on a platform fixed to the support structure of the test cell. Two sharp metallic holders rigidly supported each end of the cylinder through small openings in the frames of the optical windows. These holders were fixed to the actuators located outside the side walls of the test section.

The electromagnetic actuators may be driven by sinusoidal, random, or transient signals and therefore suitable for either open loop or feedback control. The useful frequency range of the actuator is 1–200 Hz. The maximum amplitude in this

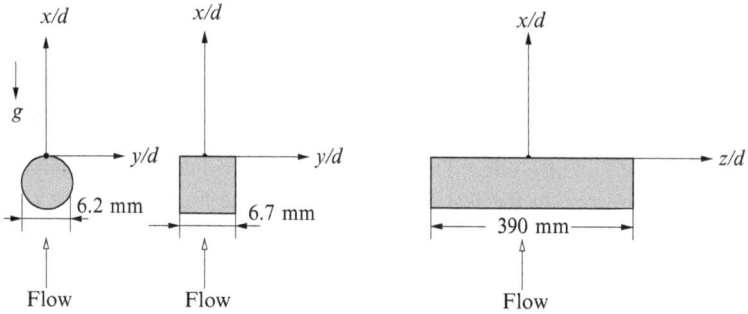

Fig. 2.3 Coordinate system for flow past a heated circular and square cylinder

Fig. 2.4 Schematic drawing of the oscillation geometry for circular and square cylinders. The symbol "a" is the amplitude of oscillation, and the *interior arrow* indicates the direction of motion of the cylinder

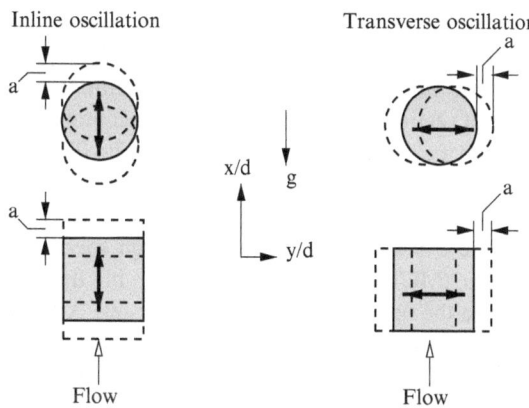

frequency range under unloaded condition is 1.7 mm, corresponding to approximately 25% of the cylinder size. Figure 2.4 shows the schematic drawing of the oscillation geometry for inline and transverse oscillation of both circular and square cylinders. The dotted line is the extreme position of the cylinders with "a" being the amplitude of oscillation. The arrow shows the direction of movement of the cylinder, being the same as the main flow for inline oscillations. For transverse oscillations, the cylinder oscillates normal to the main flow direction.

2.3.1 Optical Imaging of Convection Patterns

The flow field has been visualized using a monochrome schlieren technique. Light deflection in a variable refractive index field generated by temperature variation is captured in the schlieren image. The intensity distribution yields an integrated property of the thermal field over the length of the cylinder. The optics used in the present study consists of a Z-type 2-mirror schlieren system as shown in Fig. 2.1.

The optical setup comprises a continuous wave He–Ne laser light source (35 mW, 632.8 nm, *Spectra-physics*), a pair of flat and concave mirrors, a knife-edge and a viewing screen. The original laser beam is diverged by a spatial filter and falls on a flat mirror that directs it on to a concave mirror. Two f/8,160-cm focal length concave mirrors are used to collimate the light beam through the test section and then refocus it onto a knife-edge. The knife-edge is placed at the focal length of the second concave mirror. In the test region, flow past the heated cylinder results in a heated wake, and hence, density gradients prevail in the region being imaged. The density field results in a refractive index field, and hence, a light intensity distribution at the image plane is obtained. As the gradients are predominantly in the vertical direction, the knife-edge is kept horizontal. The knife-edge cuts off a part of the incident light for observing the schlieren effect in the form of an intensity contrast. The initial percent cutoff is adjusted in such a way that the thermal field is imaged with good contrast but without saturating the camera. For imaging of flow field, two optical windows (BK-7, 60-mm diameter, 5-mm thickness, $\lambda/4$ flatness, *Standa*) have been used to facilitate passage of the laser beam through the test section. They are mounted on accurately cut seats in Plexiglas frames with Teflon tape sealing the circumference. The Plexiglas frames fit in the 10×8 cm size rectangular slots, cut on opposite sides of the test section. Considerable care has been taken to ensure parallelism between two optical windows placed at the entrance and the exit of the laser beam. A high-speed CMOS sensor-based monochrome camera (MC1302, *Mikrotron*) with 8-bit dynamic resolution has been used to capture the sequence of schlieren images. A high knife-edge cutoff increases the sensitivity of schlieren apparatus in the sense that small density gradients in the flow can be detected. However, this setting diminishes the amount of light reaching to the camera. Short exposure time in high-speed imaging requires high light intensity to cover the full dynamic range. Hence, to balance exposure time for light intensity with high-speed imaging, an optimum is arrived at between knife-edge cutoff and frame speed to cover the full dynamic range of the camera. After various trials, the flow field was recorded at a speed of 250 frames per second and a spatial resolution of 512×512 pixels. This frame speed is well above the Nyquist criterion to capture the vortex-shedding frequency and resolve the events occurring in the development of flow structures. The camera is connected to a PC-based image acquisition system through a 64-bit frame grabber card (X64-CL, *Coreco Imaging*) and video capture software.

In the schlieren technique, the light beam closer to the cylinder is refracted due to gradients in temperature, thus generating an intensity distribution on the imaging plane. Away from the cylinder surface, temperature gradients are small, and the object beam passes with a small deflection through the test medium. Hence, if sufficiently long phase objects are available, the light beam away from the cylinder can act as a reference beam with which the test beam closer to the cylinder can interfere. Interference of the two portions of the light beam results in interferometric fringes. Fringes are the lines of constant optical path length. If the index of refraction varies only in a plane perpendicular to the beam, fringes are also contours of constant index of refraction. In ideal gas at constant pressure and constant gas composition, the fringes are lines of constant temperature averaged in the viewing direction within the test cell. Thus, schlieren interferograms provide additional information in the form of isotherms in the wake.

Errors in optical instrumentation are associated with misalignment of the apparatus with respect to the light beam and imperfection in the optical components. The noise in the optical signal during any stage of the experiment can be due to internal and external disturbances such as floor vibration and external convection patterns including air currents from fans used in cooling electronic components. This is particularly true for schlieren imaging of low velocity airflow. Experiments were conducted under quiet conditions especially during recording schlieren images. The repeatability of the data presented here was ensured from many experiments performed under practically identical test conditions.

2.4 Influence of Buoyancy

This section presents schlieren images of flow past circular and square cylinders maintained at a temperature above the ambient. The wake characteristics as a function of cylinder temperature have been investigated. At low Reynolds numbers, thermal buoyancy forces have a significant influence on the flow field. Therefore, the main focus of the present investigation is understanding the mechanisms involved in the heat-induced changes of the vortex structures under the influence of buoyancy. Numerical studies on the subject have focused on a circular cylinder. This geometry is essentially different from the square cylinder whose corners serve as fixed points of separation. Therefore, attention is directed toward the comparison of vortex-shedding mechanism of a heated circular cylinder with that of the square.

Air properties that are affected by heating are density and viscosity. To a first approximation, density reduction scales linearly with temperature, while viscosity increases as the square root of absolute temperature. For a 50 °C change in temperature, viscosity of air increases by about 8% while density decreases by 16.5%. The former affects viscous forces, while the latter generates buoyancy forces. Post separation, viscous forces are of secondary importance, and density changes are expected to have stronger influence. Experiments have been conducted at Reynolds numbers $Re = U_\infty d / v$ of 50–110 where U_∞ is the free stream velocity, d is diameter (circular) or the edge of the cylinder (square), and v is the kinematic viscosity of the fluid. The lowest flow velocity is 0.14 m/s, corresponding to a Reynolds number of 56. Fluid properties in dimensionless parameters have been evaluated at free stream conditions. The mixed convection parameter, namely, Richardson number, is defined as

$$Ri = \frac{g\beta\Delta T d}{U^2} \qquad (2.1)$$

with an appropriate length scale d and characteristic temperature difference ΔT. The cylinder is maintained at constant temperature above the ambient temperature over a Richardson number range of 0.025–0.314. The lower limit of Richardson number represents near-forced flow conditions, and the upper limit falls in the mixed convection regime. The largest temperature difference employed in the present study is 50 °C.

The physical picture of vertical flow around a heated horizontal cylinder can be visualized as follows. The fluid particles accelerate in the boundary layers around the cylinder. Increase in kinetic energy delays the point of separation, thus raising the average base pressure. Consequently, the vortices formed in the wake are weakened by heating, ultimately resulting in complete suppression of vortex shedding. At elevated cylinder temperatures, vortices are weak and result in diminishing cross-stream transport. It is a minimum when vortex shedding is absent. The progression toward the state of no vortex shedding is, however, not monotonic. At mild heating levels, the acceleration of the fluid particles in the shear layers enhances vortex strength and moderately increases cross-stream diffusion. These trends are further discussed in this chapter.

2.4.1 Instantaneous Schlieren Images

The changes in the organized wake structures with respect to their shape, size, and time-dependent movement are readily perceived from instantaneous schlieren images before the wake degenerates into a steady plume [2, 12, 28]. Reynolds numbers of $Re = 110$ for the circular cylinder and $Re = 109$ for the square are shown in Figs. 2.5 and 2.6. The effect of progressively increasing Richardson number is examined. Each time sequence has eight images in a row, separated by a time interval of 1/8th the time period of vortex shedding.

Figure 2.5a presents instantaneous images at a cylinder surface temperature of 40 °C ($Ri = 0.052$). The heated wake zones, i.e. the bright zones above the cylinder extend to a downstream distance of about $x/d = 6.0$. The near-field region ($x/d < 1$) close to the cylinder shows small variation with time and is practically stationary during the complete vortex-shedding cycle. The far-field region shows a stronger time-dependence. Thus, the base region of the wake shows very low levels of velocity and temperature fluctuations. The growth of the heated shear layer on both sides of the cylinder takes place asymmetrically at different phases of the vortex-shedding cycle. The instability of the growing shear layers results in the shedding of two alternate rows of vortices from the opposite side of the cylinder. Figure 2.5b shows instantaneous images of the cylinder wake at a surface temperature of 60 °C ($Ri = 0.104$). Vortex shedding at the higher Richardson number is quite distinct. Clear fringes, superposed on the schlieren patterns, are visible within the near wake indicating the presence of temperature variation inside the vortex. Figure 2.5c shows the time sequence of schlieren interferograms behind the circular cylinder at a surface temperature of 75 °C ($Ri = 0.140$). Regular vortex shedding along with a

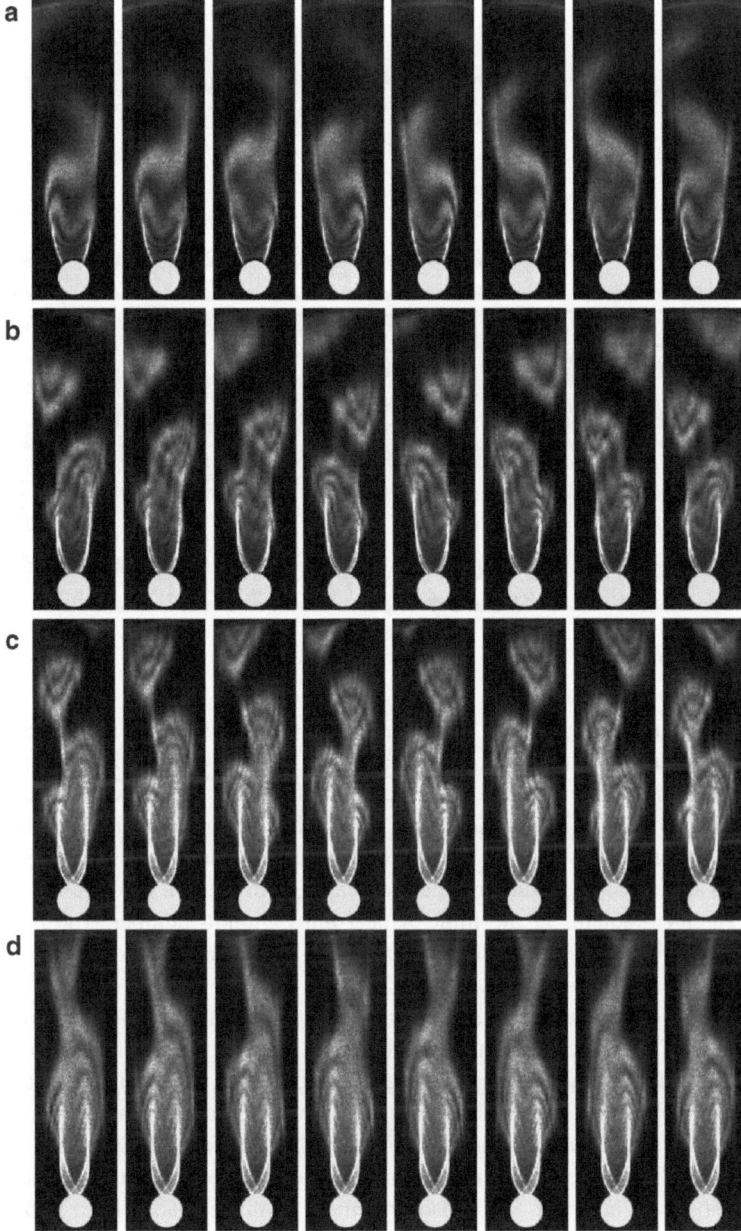

Fig. 2.5 Instantaneous schlieren images (**a**)–(**f**) for a circular cylinder separated by a time interval of one-eighth of the time period of vortex shedding at $Re = 110$ for different heating levels, that is, Richardson number. (**a**) $Ri = 0.052$, (**b**) $Ri = 0.104$, (**c**) $Ri = 0.140$, (**d**) $Ri = 0.145$, (**e**) $Ri = 0.150$, and (**f**) $Ri = 0.157$. For $Ri \geq 0.157$, images show steady state

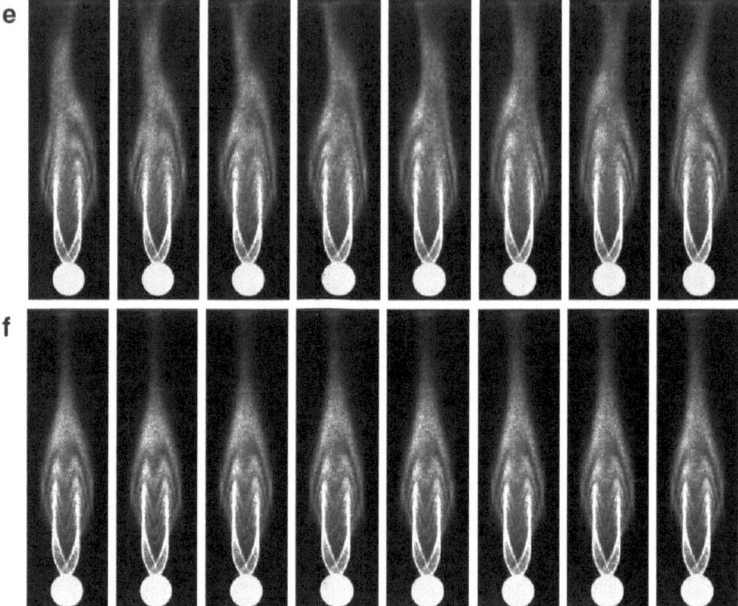

Fig. 2.5 (continued)

higher number of fringes when compared to $Ri = 0.104$ is seen. This indicates that vortex is not an isothermal packet of fluid; rather it is the recirculation bubble with a temperature distribution within. With increasing Richardson number, the increase in the number of fringes is a consequence of a higher overall temperature difference, on the one hand, and large localized temperature gradients, on the other. These gradients arise from the structure of a shear layer, with thermal gradients correlating with velocity gradients at the limit of near-unity Prandtl number. The instability mechanisms that result in the detachment of the vortex continue to be quite similar at Richardson numbers up to 0.140.

Figure 2.5d shows instantaneous images at a surface temperature of 77 °C ($Ri = 0.145$). The vortex structure starts to transform compared to the experiments at lower Richardson number. Large vortices are converted to thin elongated structures interacting from each side of the wake. The flow visualization images during one cycle of vortex shedding are shown in Fig. 2.5e for a cylinder temperature of 79 °C ($Ri = 0.150$). The vortex structure is completely altered here. The alternate shedding pattern observed at lower Richardson numbers is now replaced by a thin plume, which slowly oscillates in the transverse direction. The interferograms in the near-field region of the cylinder fluctuate in phase with the cross-stream oscillation of the thin plume region. The schlieren interferograms at a higher surface temperature of 82 °C ($Ri = 0.157$) are presented in Fig. 2.5f. The thin plume in the far-field region is now steady, aligned with the cylinder midplane. The plume is thinner compared to $Ri = 0.150$, and the wake can be termed as steady.

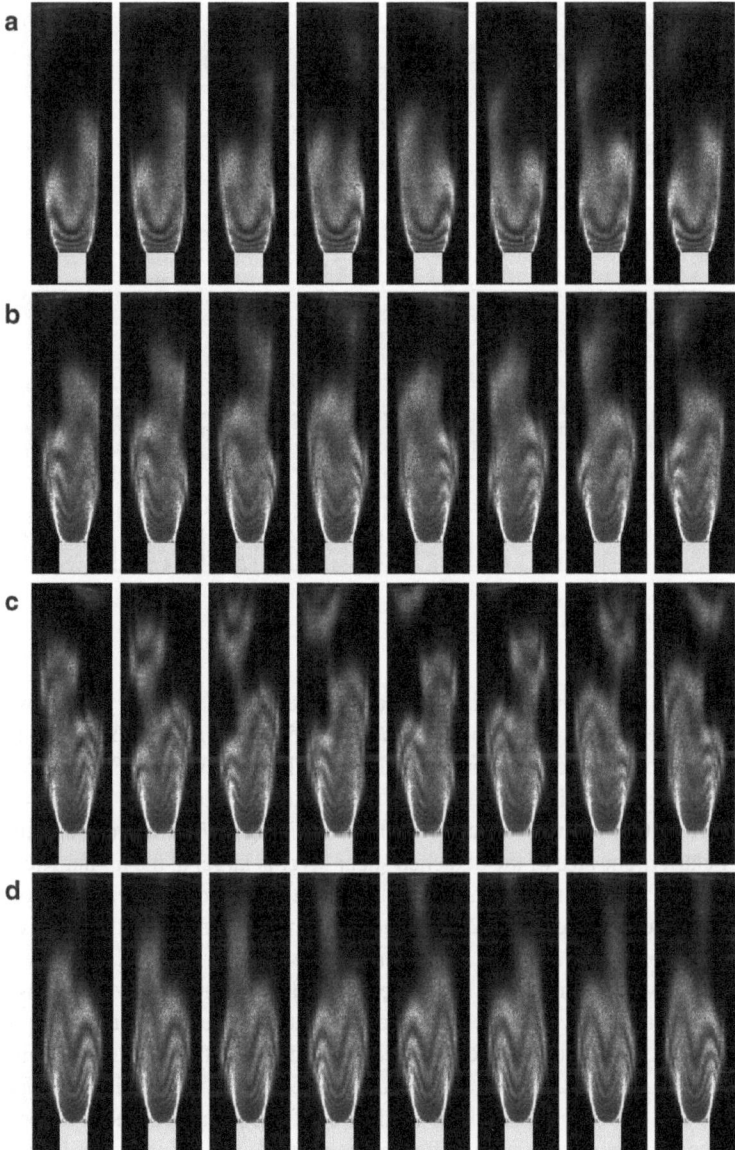

Fig. 2.6 Instantaneous schlieren images (**a**)–(**f**) for a square cylinder separated by a time interval of one-eighth of the time period of vortex shedding at $Re = 109$ for different heating levels, that is, Richardson number. (**a**) $Ri = 0.059$, (**b**) $Ri = 0.108$, (**c**) $Ri = 0.117$, (**d**) $Ri = 0.124$, (**e**) $Ri = 0.133$, and (**f**) $Ri = 0.155$. For $Ri \geq 0.155$, images show steady state

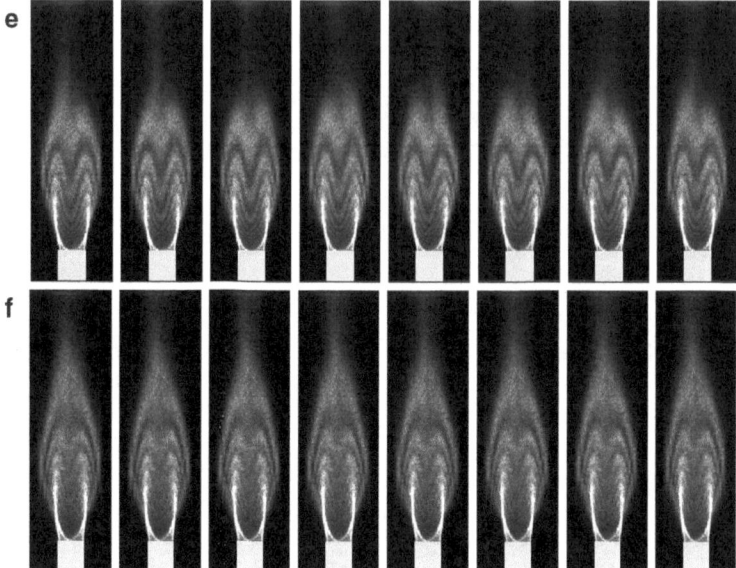

Fig. 2.6 (continued)

Figure 2.6a presents the visualization images for a square cylinder when its surface temperature is equal to 40 °C ($Ri = 0.059$). The oscillation in the far-field region with vortex shedding from opposite shear layers is visible. Figure 2.6b shows the instantaneous schlieren images at a surface temperature of 55 °C ($Ri = 0.108$). Compared to the lower Richardson number, the images show vortex shedding with greater clarity. The size of the detached shear layer is larger at the higher Richardson number. Vortex shedding here is less distinct for the square cylinder when compared to the circular cylinder (Fig. 2.5b). The schlieren interferograms at the cylinder temperature of 58 °C ($Ri = 0.117$) for the square cylinder in Fig. 2.6c show distinct and regular vortex shedding. The interferometric fringes are visible inside the detached vortices. Hence, the increase in heating level regularizes the vortex-shedding process, though mixed convection influences the interaction among the shed vortices. Figure 2.6d shows schlieren interferograms for a cylinder temperature of 60 °C ($Ri = 0.124$). The shape and size of the detached shear layer at this Richardson number is distinctly different from that at the lower Richardson number ($Ri = 0.117$). The detached shear layer arising from the vortex-shedding process is elongated at the higher Richardson number. The instantaneous images at higher cylinder temperature of 63 °C ($Ri = 0.133$) in Fig. 2.6e indicate mild unsteadiness of the shear layer. Figure 2.6f shows the schlieren images at cylinder temperature of 70 °C ($Ri = 0.155$). Here, two shear layers have merged into one leading to a single steady plume at the center plane of the cylinder. The schlieren interferograms in the near-field region are also steady in time, indicating complete steadiness of the wake. However, the plume extends farther downstream for the higher Richardson number ($Ri = 0.155$) when compared to the lower ($Ri = 0.133$).

The schlieren interferograms of both circular and square cylinder show similar vortex structures as a function of heating level, that is, Richardson number. The mechanism for suppression of vortex shedding can be identified from the schlieren images as follows: Heating reduces fluid density, and a small increase in cylinder surface temperature accelerates fluid particles and feeds vorticity to the shear layer, which leads to regular vortex shedding. Thus, the first effect of cylinder heating is to increase the Strouhal number with Richardson number. With further increase in heating level, the diffusion of thermal energy in the near wake creates stabilizing buoyant forces with respect to the fluid at ambient temperature contained in the shear layer. Stratification suppresses vortex shedding when the Richardson number exceeds the critical value. Wake unsteadiness (and mixing) is now diminished, and a nearly steady plume is observed. The corresponding schlieren images have a uniform spread of light intensity over the wake region.

2.4.2 Strouhal Number: Effect of Heating

The variation of Strouhal number with Richardson number for given Reynolds numbers is presented in Fig. 2.7a, b, respectively, for circular and square cylinders. For each Reynolds number, both figures show a small initial increase in Strouhal number with Richardson number. Strouhal number is higher at higher Reynolds numbers. In the Reynolds number range considered, Strouhal number follows an identical trend with respect to Reynolds number for the heated cylinder as for an unheated cylinder [18, 25, 29, 30]. Hence, it is seen that the effect of Reynolds number on Strouhal number is not affected by heating till the critical point is reached. The increase in Strouhal number with increase in heating level, that is, Richardson number, takes place for both circular and square cylinder (Fig. 2.7a, b) till a critical point is reached. Beyond the critical Richardson number, vortex shedding stops abruptly, and Strouhal number falls to zero. Earlier numerical studies [5, 11, 24, 27] have shown an increase in Strouhal number with increase in Richardson number, followed by a singular decay to zero at the critical Richardson number. These studies pertain to a fixed Reynolds number $Re = 100$ for a circular cylinder. The present experimental study confirms this trend for both circular and square cylinders for Reynolds numbers around 100.

2.4.3 Critical Richardson Number: Effect of Reynolds Number

The critical Richardson number at which suppression of vortex shedding takes place depends on Reynolds number. It is higher at higher Reynolds numbers as depicted

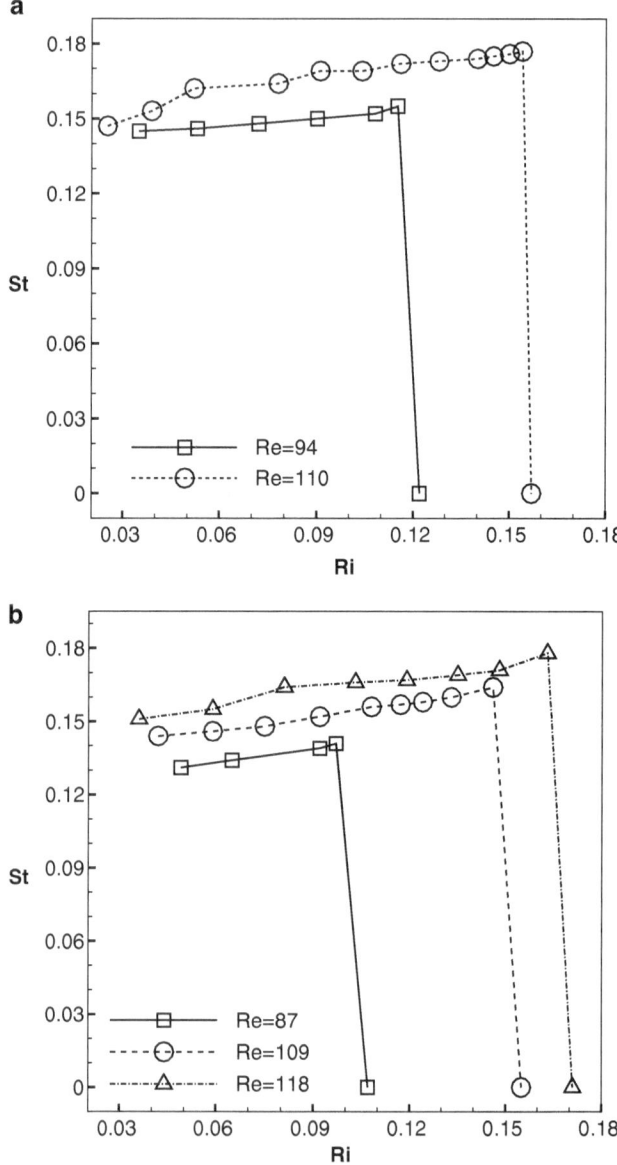

Fig. 2.7 Variation of Strouhal number with Richardson number for a (**a**) circular cylinder and (**b**) square cylinder

from the points of zero Strouhal number in Fig. 2.7a, b. Table 2.1 presents critical Richardson number data as a function of Reynolds number for circular and square cylinders. The disappearance of vortex shedding takes place at $Ri = 0.122$ for $Re = 94$ and at $Ri = 0.157$ for $Re = 110$ for a circular cylinder. The critical Ri values for

Table 2.1 Critical Richardson number and critical Reynolds number (Re_w, Re_{film}, and Re_{eff}) calculated using kinematic viscosity of air at T_w, T_{film}, and T_{eff} at different free stream velocities for circular and square cylinders

S.N.	Cylinder geometry	U_∞ (m/s)	Re_∞	Ri_{crit}	$T_\infty °C$	$T_w °C$	T^*	$T_{film} °C$	$T_{eff} °C$	Re_w	Re_{film}	Re_{eff}
1.	Circular	0.23	94	0.122	21	54	1.11	37.5	30.2	78	85	89
2.	Circular	0.27	110	0.157	21	82	1.21	51.5	38.1	79	92	100
3.	Square	0.20	87	0.107	23	43	1.07	33	28.6	78	82	84
4.	Square	0.22	94	0.121	26	54	1.09	40	33.8	80	87	90
5.	Square	0.24	103	0.140	26	65	1.13	45.5	36.9	82	92	96
6.	Square	0.25	109	0.155	23	70	1.16	46.5	36.2	84	95	101
7.	Square	0.27	118	0.171	23	85	1.21	54	40.4	84	99	107

a square cylinder are equal to 0.107, 0.121, 0.140, 0.155, and 0.171 for $Re = 87$, 94, 103, 109, and 118, respectively. The best line passing through this data is of the form

$$Ri = -7.58 \times 10^{-2} + 2.1 \times 10^{-3} Re$$

with a regression coefficient of 99.7%.

The critical Richardson numbers for a square cylinder are quite close to those of a circular cylinder at comparable Reynolds numbers (Table 2.1). This indicates that both circular and square cylinders share factors that are similar for suppression of vortex shedding. From numerical simulation, a delay in the point of separation over a circular cylinder was reported due to heating [20]. For a square cylinder, the separation point is fixed at the corners of the cylinder. Since experiments show the critical Richardson number for circular and square cylinders to be comparable, one can conclude that shift in the point of separation is not primarily responsible for suppression of vortex shedding. It has been attributed to the breakdown of the vortex street due to acceleration of velocity in the wake and not a shift in the points of separation [22].

2.4.4 Vortex Formation Length

The streamwise location of a peak in RMS velocity along the centerline of a cylinder corresponds to the initiation of vortex shedding and hence the vortex formation length. Konstantinidis et al. [15] have presented the vortex formation length from the locations of peaks in the velocity fluctuations along the cylinder centerline. It can be assumed that the light intensity fluctuation in schlieren images is directly related to the temperature fluctuation and indirectly to velocity fluctuations. Along the centerline, the average light intensity is small and the signal-to-noise ratio is poor. Hence, the vortex formation length is based on the intensity fluctuations in schlieren images at a small, offset location of $y/d = 0.25$ from the cylinder centerline.

Table 2.2 Vortex formation length as a function of Reynolds number and surface temperature for a circular cylinder. Here, NS (*no shedding*) indicates suppression of vortex shedding

Cylinder surface temperature, °C	L_f/d $Re = 94$	L_f/d $Re = 110$	Cylinder surface temperature, °C	L_f/d $Re = 110$
30	2.0	2.0	60	3.7
35	2.4	2.5	70	3.8
40	2.5	2.9	75	3.9
45	3.0	3.6	77	4.0
50	3.6	3.7	79	4.0
52	3.6	–	81	4.0
54	NS	–	82	NS

Table 2.3 Vortex formation length as a function of Reynolds number and surface temperature for a square cylinder. Here, NS indicates suppression of vortex shedding

Cylinder surface temperature, °C	L_f/d $Re = 87$	L_f/d $Re = 109$	L_f/d $Re = 118$	Cylinder surface temperature, °C	L_f/d $Re = 109$	L_f/d $Re = 118$
35	1.8	1.8	1.8	59	–	3.1
40	2.7	2.4	–	60	3.9	–
41	2.9	–	–	63	3.9	–
43	NS	–	2.4	65	3.9	3.6
45	–	2.6	–	67	3.9	–
47	NS	–	–	70	NS	–
50	NS	3.1	–	71	–	3.9
51	–	–	2.9	76	NS	4.3
55	–	3.5	–	82	–	4.3
58	–	3.6	–	85	–	NS

The dimensionless vortex formation lengths, (L_f/d) obtained under experimental conditions, are compared in Tables 2.2 and 2.3 for the circular and square cylinders, respectively. The vortex formation length is a function of Richardson number. For a low Richardson number, corresponding to surface temperatures of 30 °C and 35 °C, the vortex formation length is found to be 2.0 and 1.8 for the circular and square cylinders, respectively. Kim et al. [13] and Konstantinidis et al. [15] reported L_f/d to be in the range of 1.5–2.3 for a circular cylinder. These are comparable to the values obtained using schlieren imaging. For both cylinders, the formation length increases with Richardson number and approaches an asymptotic value, appropriate for each Reynolds number. The asymptotic formation lengths (L_f/d) are 2.9, 3.9, and 4.3 for $Re = 87$, 109, and 118 for the square cylinder. For circular cylinder, the formation length values are 3.6 and 4.0 at $Re = 94$ and $Re = 110$, respectively.

The critical Richardson number at which suppression of vortex shedding takes place is higher at higher Reynolds numbers. Simultaneously, the Strouhal number prior to suppression is also higher. Hence, the formation length seems to be correlated to both critical Richardson number and Strouhal number. An increase in fluid speed in a highly buoyant flow field can explain longer distances traversed (increase in formation length, L_f) as well as a lower time period (increase in Strouhal number upto the critical point). The increase in size of the vortex formation region with heating level can be related to the delay in transition of the separating shear layer. Buoyancy has the effect of diffusing the vorticity content inside the shear layer and the buoyant plume reduces the interaction between opposite shear layers formed on each side of the cylinder. Beyond the critical Richardson number, vortex shedding is suppressed and the vortex formation length is no longer defined.

2.4.5   Distribution of Light-Intensity Fluctuations

Since light measurements are inertia free, time-dependent fluctuations in light intensity from schlieren interferograms are essentially due to the temperature gradient (and hence, temperature) fluctuations. As the working fluid is air with a Prandtl number of the order of unity, these can further be correlated to velocity fluctuations. The correlation is meaningful outside the recirculation zone, where advection effects are much larger than diffusion. The influence of heating on wake fluctuations of circular and square cylinders is of interest in the present section. The RMS profiles of the fluctuating light intensity (with mean removed) in the streamwise direction (x/d) at various transverse locations (y/d) are presented in Figs. 2.8 and 2.9 for the circular and square cylinders, respectively. The corresponding Reynolds numbers for these cases are 110 and 109. Figures 2.8 and 2.9 indicate strong fluctuations in case of vortex shedding, which diminish at higher heating levels due to vortex suppression. For both cylinders, the intensity of fluctuations is small in the near-field region $(x/d \leq 2)$, very close to the cylinder surface. The instantaneous images in Figs. 2.5 and 2.6 also show steady fringes in the near-field region, supporting the trends in RMS intensity of Figs. 2.8 and 2.9. The jump in RMS intensity is due to inception of vortex shedding. Therefore, RMS values are higher in the far-field region where the flow is driven by sustained vortex shedding. The unsteadiness of this region was also recorded in the schlieren images of Figs. 2.5 and 2.6. The RMS light intensity increases from a small value to a maximum in the streamwise direction and subsequently reduces with distance. The streamwise location of the peak value of RMS intensity fluctuation from the surface of the cylinder is the vortex formation length. The fluctuation in light intensity is higher inside the shear layers $(y/d = 0.45)$ and reduces toward the cylinder centerline $(y/d = 0)$. These trends are supported by the RMS profiles of light intensity fluctuations in the transverse direction (y/d) for given streamwise locations.

The maximum RMS light intensities are functions of the Richardson number. At lower Richardson numbers ($Ri = 0.052$ for circular and $Ri = 0.059$ for the square), the light intensity fluctuations are small. For a moderate increase in the cylinder temperature, the vortex strength is enhanced, being accompanied by regular and strong vortex shedding. Therefore, the maximum RMS intensity increases initially with Richardson number. At elevated cylinder temperatures, the shed vortices become weak and interactions between the vortices from the opposite shear layers

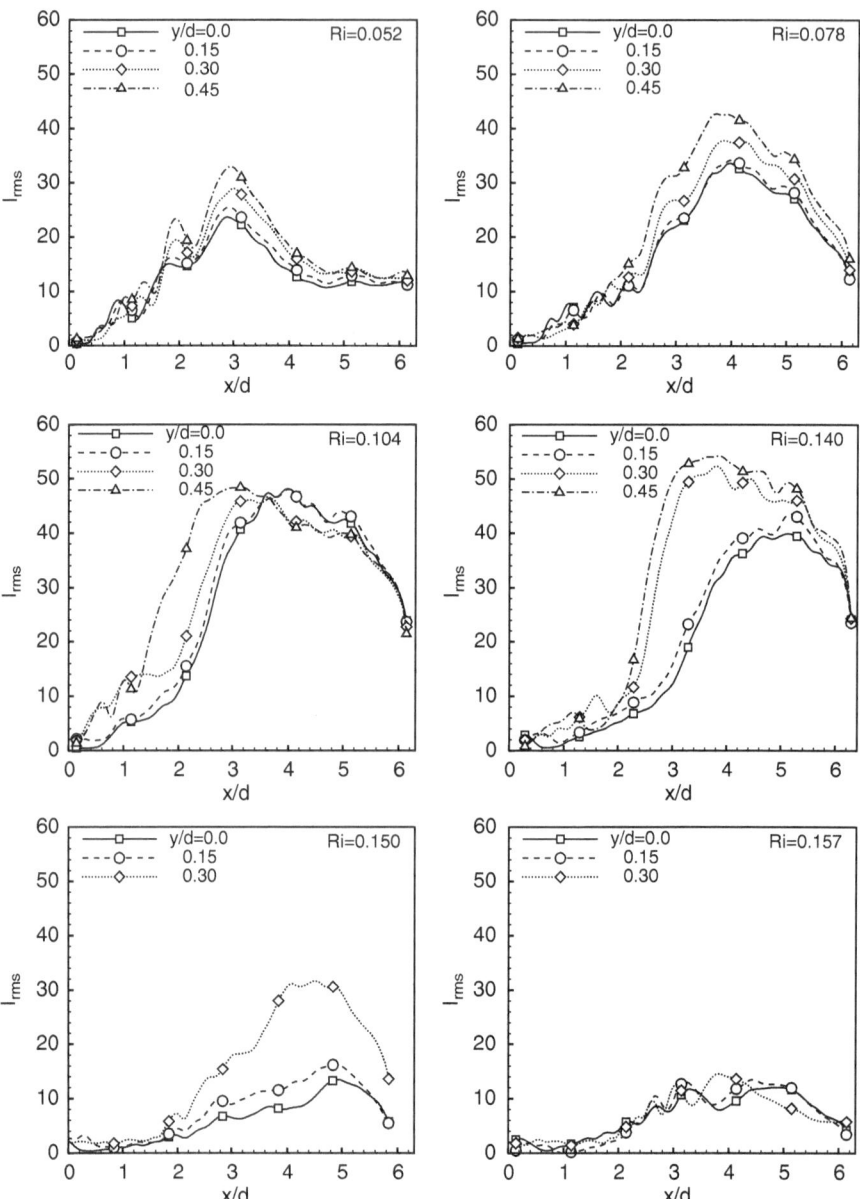

Fig. 2.8 Circular cylinder: evolution of the RMS intensity (I_{rms}) in the streamwise direction at various transverse locations, ($y/d = 0.0$, 0.15, 0.30 and 0.45), $Re = 110$. Dependence on Richardson number is shown

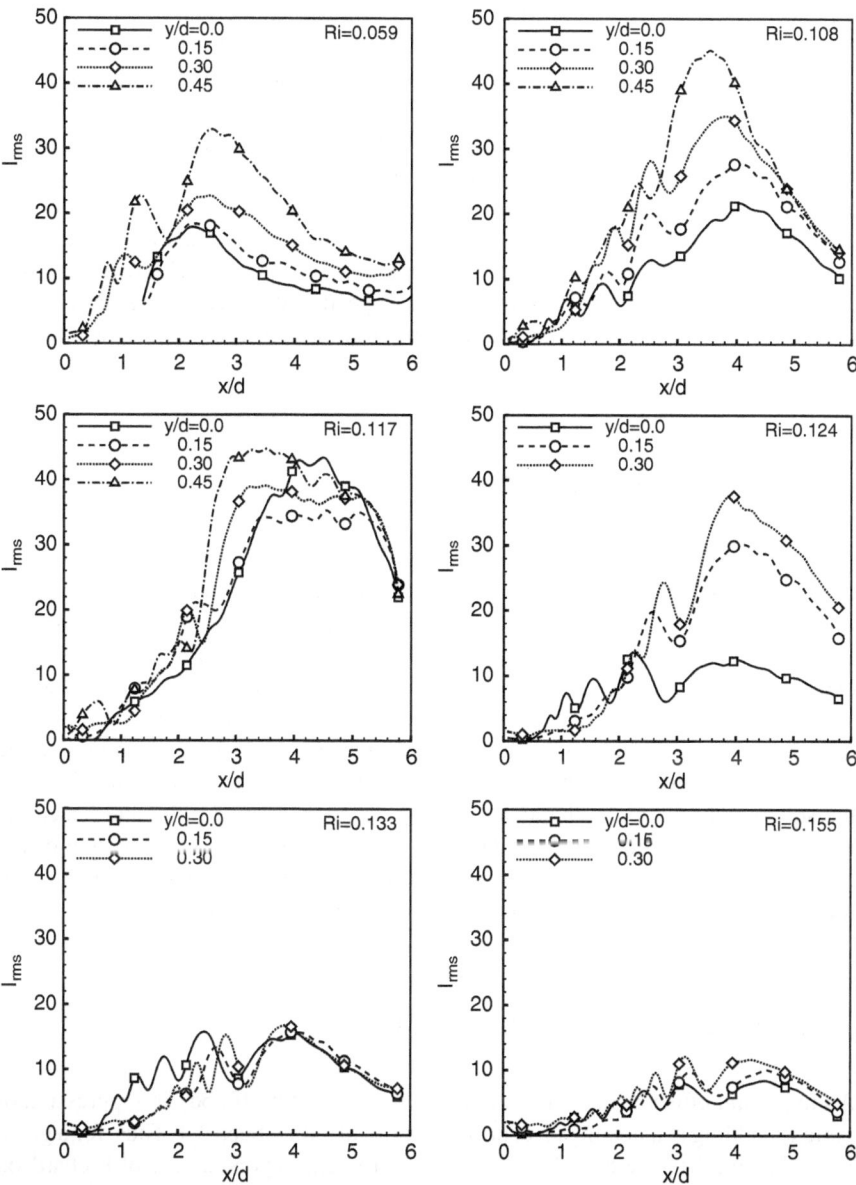

Fig. 2.9 Square cylinder: evolution of the RMS intensity (I_{rms}) in the streamwise direction at various transverse locations, ($y/d = 0.0$, 0.15, 0.30 and 0.45), $Re = 109$. Dependence on Richardson number is shown

are also weakened. Therefore, RMS intensity is lower at a Richardson number just below the critical. At the highest Richardson number ($Ri = 0.157$ for circular and 0.155 for the square), the RMS values of light fluctuations are insignificant due to the complete suppression of vortex shedding.

The relative difference in magnitude of the RMS intensity at various transverse locations (y/d) is an indicator of the size and shape of the detached vortices. In the vortex-shedding regime, the RMS intensity increases in the transverse direction away from the centerline. Factors that determine the flow fluctuation variations in the wake are (a) the average size of the shed vortex, (b) interaction between opposite shear layer vortices, and (c) the average fluid speed in the shear layers. Figures 2.8 and 2.9 can be reinterpreted to examine the transverse variation of light intensity fluctuations at various streamwise locations.

2.5   Effect of Cylinder Orientation

The effect of the orientation of a square cylinder on the wake properties has been studied by various authors [3, 7, 14, 29]. Flow past a heated horizontal square cylinder placed at an angle of incidence with respect to the main flow in the vertical direction is investigated in the present study. The sharp corners of the square cylinder fix the points of separation. However, the surface temperatures influence the evolution of the boundary layer beyond the separation points. The effect of fluid heating on vorticity generation via buoyancy controls the stability of the shear layer and hence the inception of vortex shedding. The angle of incidence of the square cylinder alters the axial location of the separation point and hence the downstream interaction between the two shear layers located on each side of the cylinder centerline. The present study investigates the wake characteristics under the combined influence of buoyancy and the angle of incidence [27].

2.5.1   Suppression of Vortex Shedding

An important effect of buoyancy on flow past a heated bluff body is suppression of vortex shedding at a heating level beyond a critical value, Sect. 2.4. Thus, for a given Reynolds number, the Strouhal number falls to zero beyond a critical Richardson number. This result has been discussed in the literature for square and circular cylinders, from computations as well as experiments described in Sect. 2.4. In the present section, the influence of cylinder inclination with respect to the main flow on suppression of vortex shedding is reported. Vortex shedding is clearly present when the spectra of light intensity fluctuations in the near wake of the cylinder show one or more clear peaks. A broadband low amplitude spectrum without isolated peaks is taken as an indication of suppression of the vortex-shedding phenomenon.

Strouhal number variations with respect to the Richardson number at Reynolds numbers of 56, 87, and 100 are presented in Fig. 2.10a–c, respectively, for various

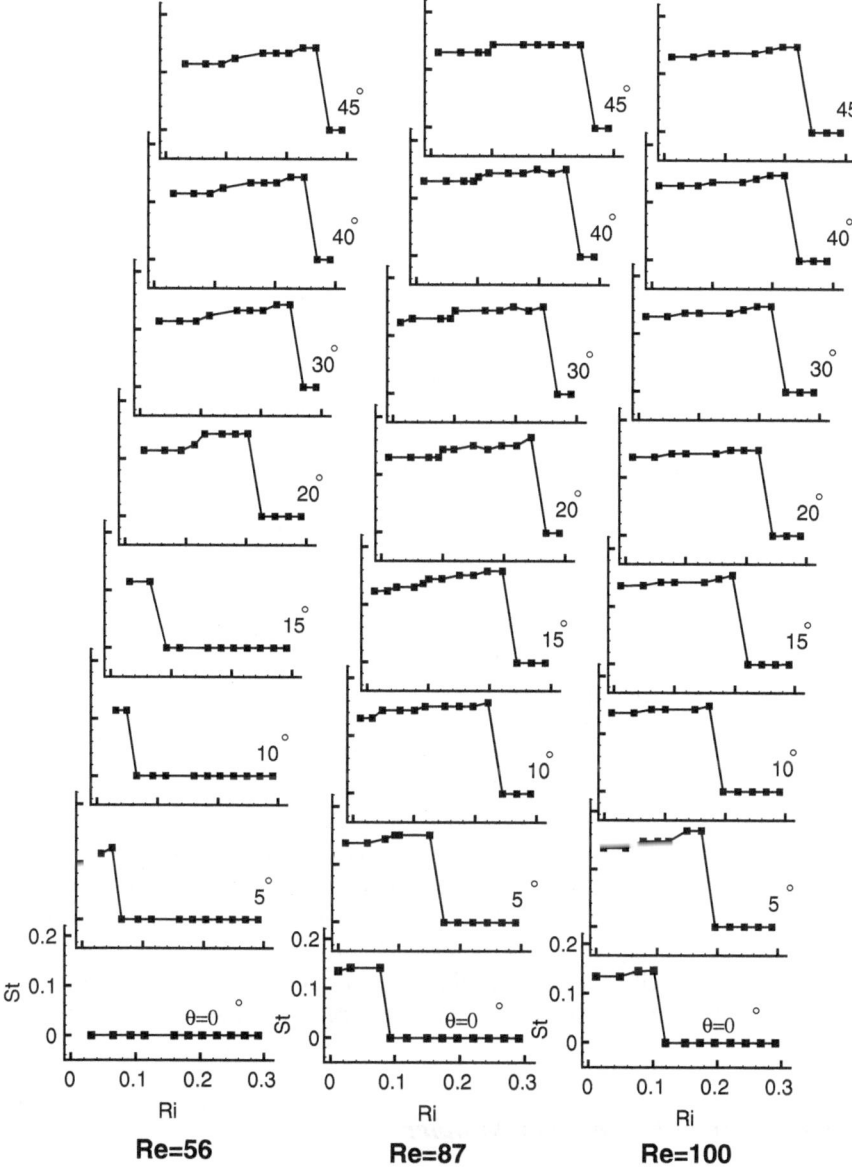

Fig. 2.10 Strouhal number variation as a function of Richardson number and cylinder orientation for different Reynolds number, that is, $Re = 56$, 87, and 100

incidence angles. At all Reynolds numbers, Fig. 2.10 shows a small initial increase in the Strouhal number with Richardson number till vortex shedding is suppressed. At this stage, a clear peak in the light intensity spectrum was not to be seen, and the Strouhal number is taken to be zero. At a Reynolds number of 56 and

Table 2.4 Critical
Richardson number as a
function of incidence angle at
various Reynolds numbers

Angle of incidence (θ) (deg)	Critical Richardson number		
	$Re = 56$	$Re = 87$	$Re = 100$
0	0.031	0.107	0.131
5	0.065	0.182	0.203
10	0.065	0.248	0.203
15	0.092	0.248	0.226
20	0.226	0.270	0.248
30	0.270	0.270	0.248
40	0.270	0.270	0.248
45	0.270	0.270	0.248

an incidence angle of $0°$, vortex shedding is absent at all Richardson numbers
(Fig. 2.10a). The overall flow is stable, and the corresponding Strouhal number is
zero. Regular vortex shedding and hence a nonzero Strouhal number are noted at
higher angles (5–45°) for $Re = 56$ (Fig. 2.10a). At Reynolds numbers of 87 and 100
and nearzero heating, regular vortex shedding is resumed for all incidence angles,
and a non-zero Strouhal number is obtained (Fig. 2.10b, c). At these Reynolds
numbers, vortex shedding is suppressed at higher levels of heating, that is, higher
Richardson numbers. The overall conclusion to emerge from these experiments is
that cylinder orientation away from $0°$ destabilizes the shear layer and promotes
instability, while heating (specifically, buoyancy) ultimately stabilizes the flow and
suppresses vortex shedding.

Data in Fig. 2.10 can be interpreted in the following manner. An increase in
the cylinder angle draws vortices from each side of the cylinder due to pressure
asymmetry, promoting vortex shedding and an increase in the Strouhal number.
When the cylinder is heated, the fluid particles in the shear layer accelerate in
the buoyancy field, vorticity generation is enhanced, and there is a possibility of
early shedding and hence a further increase in Strouhal number. In contrast, heating
creates buoyancy forces that tend to compensate for the momentum deficit in the
wake. This factor delays vortex shedding, and at a certain level of heating, vortex
shedding is completely suppressed. At this stage, heating has substantially overcome
the effect of cylinder orientation as well as fluid acceleration due to buoyancy.

2.5.2   Critical Richardson Number

Table 2.4 compiles the critical Richardson number data as a function of the Reynolds
number and the incidence angle. The critical Richardson number is the one for
which the Strouhal number has abruptly decreased to near-zero values. It was
identified with the following approach. The reference spectrum was taken as the
one at the smallest Richardson number and zero angle of incidence where a single
peak was clearly visible in the vortex-shedding regime. For nonzero angles of
incidence as well as for increasing values of Richardson number, multiple peaks

were seen the light intensity spectra. However, the peak closest to the one in
the reference spectrum was used for the determination of the Strouhal number.
In select experiments close to transition, the schlieren images did not reveal the
Karman vortex-shedding pattern though a peak in the spectrum was discernible
(e.g., $Re = 100$, $Ri = 0.182$, $\theta = 10°$). In such instances, the spectral information of
frequency was used to determine the Strouhal number.

For a given Reynolds number, Table 2.4 shows that increasing the incidence angle
increases the critical Richardson number up to a limit of around $20°$. The critical
Richardson number is less sensitive to cylinder orientation beyond the incidence
angle of $20°$. For $Re = 56$, one observes flow separation for all configurations,
but vortex shedding is seen only for non-zero angles of the cylinder inclination. In
the absence of shedding, the critical Richardson number refers to the heating level
needed to suppress flow separation. This value is relatively modest in comparison
to the critical that is required to suppress vortex shedding, the latter being a
dynamically stronger phenomenon. The sudden jump in Ri (critical), going from
$\theta = 15$ to $20°$ is thus related to the onset of vortex shedding in the base flow, from
a state of flow separation.

At lower incidence angles up to $10°$, the critical Richardson number monotoni-
cally increases with the Reynolds number of the approach flow. However, at higher
incidence angles (example, $\theta = 20°$) the critical Richardson number increases from
$Re = 56$ to $Re = 87$ while it drops to a lower value at $Re = 100$. This trend indicates
that the nature of vortex shedding and the interaction between the opposite shear
layers are influenced differently by buoyancy at lower and higher incidence angles.
When the cylinder is misaligned with the flow, the shear layers originating from
each side of the cylinder combine and result in a net cancellation of vorticity. With
increased heating, the shear layers accelerate and carry greater vorticity content. In
addition, buoyancy tends to cancel the momentum deficit in the wake. These factors
are, thus, competing in nature in the sense that vortex cancellation and the possibility
of vortex shedding jointly become more probable at the higher Reynolds number.

2.5.3   Instantaneous Schlieren Images

The changes in the organization of the wake structure, namely, its shape, size,
and time-dependent movement, are readily perceived from the time sequence of
schlieren images for various Richardson numbers. Within a vortex-shedding cycle,
Fig. 2.11 show a time series of instantaneous schlieren images for an orientation
angle of $45°$. A Reynolds number of 100 has been considered with Richardson
numbers equal to 0.031, 0.114, 0.182, 0.203, and 0.248, respectively. The initial
image recorded on the screen is one of uniform darkness. Under heated conditions,
bright streaks are created around the cylinder. The brightest light streaks are to be
found in the shear layers originating from the cylinder corners. The wake as a whole
is brighter than the background indicating finite temperature gradients distributed
within.

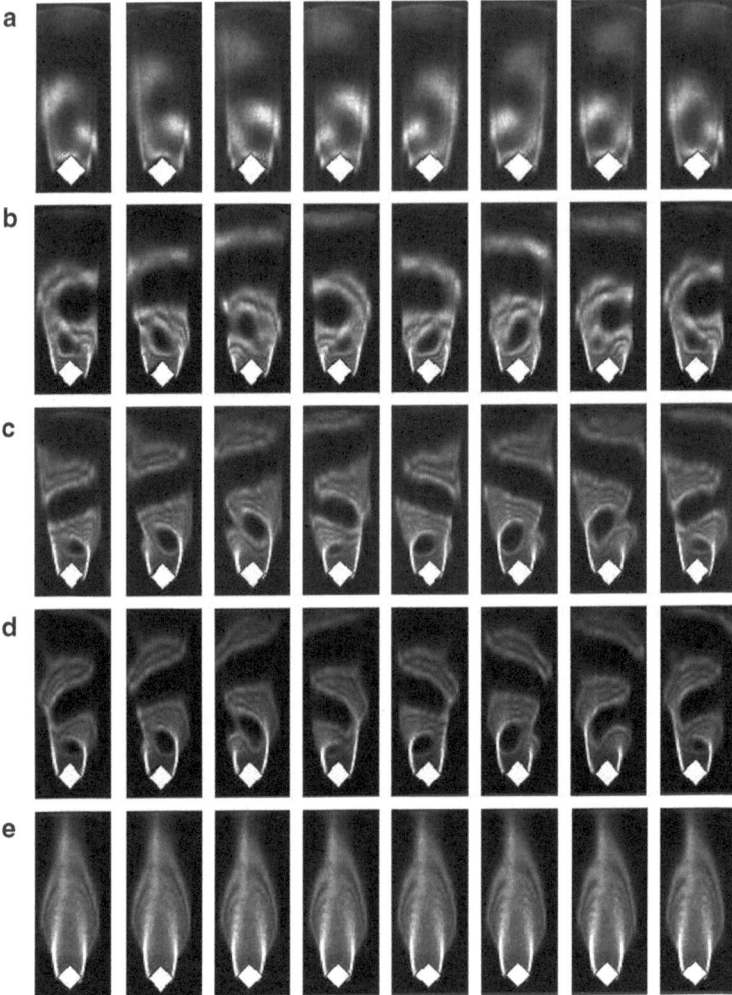

Fig. 2.11 Instantaneous schlieren images (**a**)–(**e**) for a square cylinder inclinded at an angle of $45°$ to the main flow. Images are separated by a time interval of one-eighth of the time period of vortex shedding at $Re = 100$ for different heating level, that is, Richardson number: (**a**) $Ri = 0.031$, (**b**) $Ri = 0.114$, (**c**) $Ri = 0.182$, (**d**) $Ri = 0.203$, and (**e**) $Ri = 0.248$. For $Ri \geq 0.248$, images show the wake replaced by a steady plume

Under forced flow conditions, the cylinder wake contains Karman vortices. High Richardson numbers lead to suppression of vortex shedding and create a steady buoyant plume. The vortex-shedding process is regularized with an increase in the Richardson number from 0.031 to 0.203 for the inclined cylinder. The critical Richardson number required to suppress vortex shedding increases with the cylinder angle. For example, when the orientation is $45°$, the wake is unsteady at $Ri = 0.203$

and $Re = 100$. The corresponding image for a cylinder angle of $0°$ shows the suppression of vortex shedding. Thus, an increase in the angle of incidence is seen as a destabilizing influence on the wake. The region close to the cylinder ($x/d < 1$) shows a very small variation with time and is practically stationary during the complete vortex-shedding cycle. The downstream region shows a stronger time dependence. Thus, the base region of the wake shows low levels of velocity and temperature fluctuations. The growth of the shear layer on each side of the cylinder takes place asymmetrically at different phases of the vortex-shedding cycle. Thus, for all Richardson numbers, shear layer instability results in the shedding of alternate rows of vortices from each side of the cylinder. In some images, fringe formation can be seen in the thermal wake. As discussed earlier in this chapter, these are related to interference effects between the portion of light refracted away from the region of high temperature with the undisturbed wavefront. Before the suppression of vortex shedding, clear fringes are visible in the cylinder wake. These are a measure of temperature variation inside the vortices.

With increasing Richardson number, the increase in the number of fringes is a consequence of higher overall temperature difference on one hand and large localized temperature gradients on the other. When shedding is suppressed, the unsteadiness of the base flow continues to be felt in the plume in the form of slow transverse oscillations. At the extreme value of $Ri = 0.248$, vortex shedding is halted for all cylinder orientations. The alternate shedding pattern observed at lower Richardson numbers is replaced by a thin plume. The plume is steady and aligned with the cylinder mid-plane for an angle of $0°$. For other angles, the plume is not symmetric. The asymmetric plume is related to the asymmetric location of the upstream edges of the square cylinder.

Figures 2.12–2.14 show a time series of instantaneous schlieren images for orientation angles of 0–45°. Wake patterns in the range of 45–90° are a repetition of those obtained between 0 and 45°. The data in these figures is for a Reynolds number of 100 and Richardson numbers of 0.031, 0.182, and 0.248, respectively.

2.6 Influence of Cylinder Oscillations

In applications involving gases (e.g., cooling of electronic equipment, gas turbine blades, and miniature heat exchangers) and liquids (such as signatures of submarines), one often encounters the need to control surface forces and maximize energy exchange between the surface and the surrounding fluid. These requirements are adequately met by modifying the wake structure. The intensity of the disturbance dictates the magnitude and temporal characteristics of forces acting on the object. For many applications, flow separation, say at a corner, creates a large wake region downstream. The evolution of the wake is composed of three fundamental features, the boundary layer, the free shear layer, and vortex, each of which develops instability as certain critical parameters are progressively increased. The wake is a consequence of such instabilities and can be identified as a field of

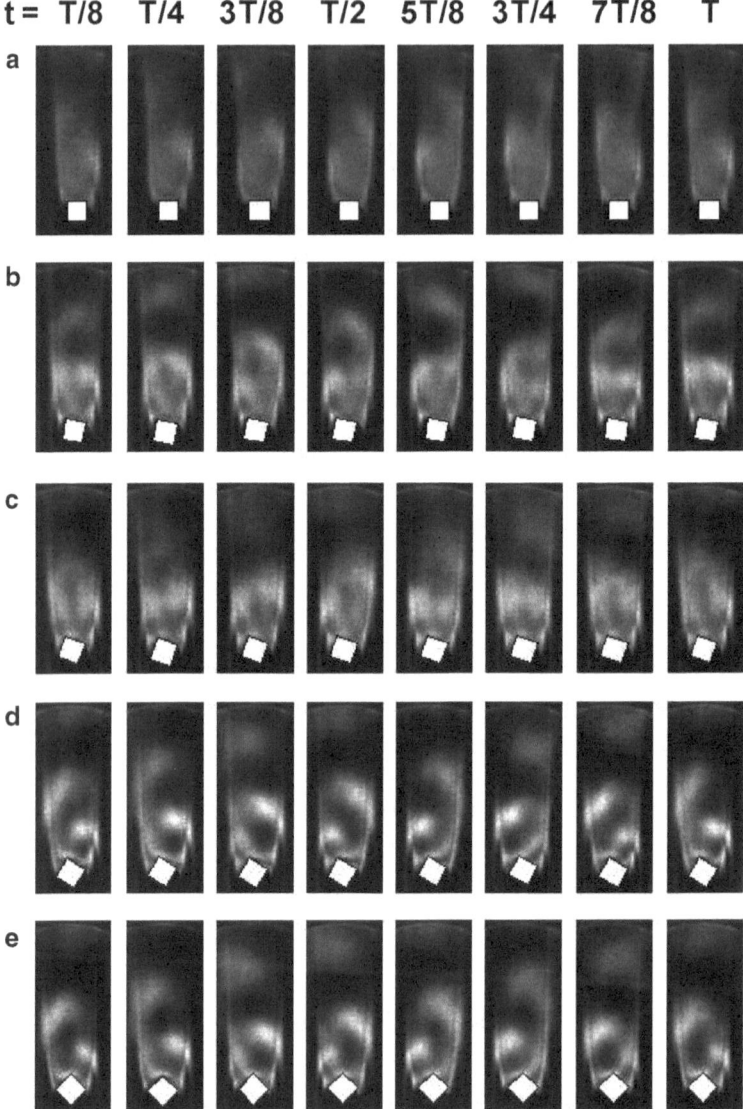

Fig. 2.12 Instantaneous schlieren images separated by a time interval of one-eighth of the time period of vortex shedding at $Re = 100$ and $Ri = 0.031$ for orientation angles of (**a**) 0, (**b**) 10, (**c**) 20, (**d**) 30, and (**e**) 45 °. The time period of vortex shedding is denoted as T

three-dimensional unsteady vorticity. The three distinct flow fields interact among each other and culminate in a complex flow structure.

The subject of flow control deals with managing forces acting on a solid surface when it is placed in a fluid environment. Two strategies available for control are

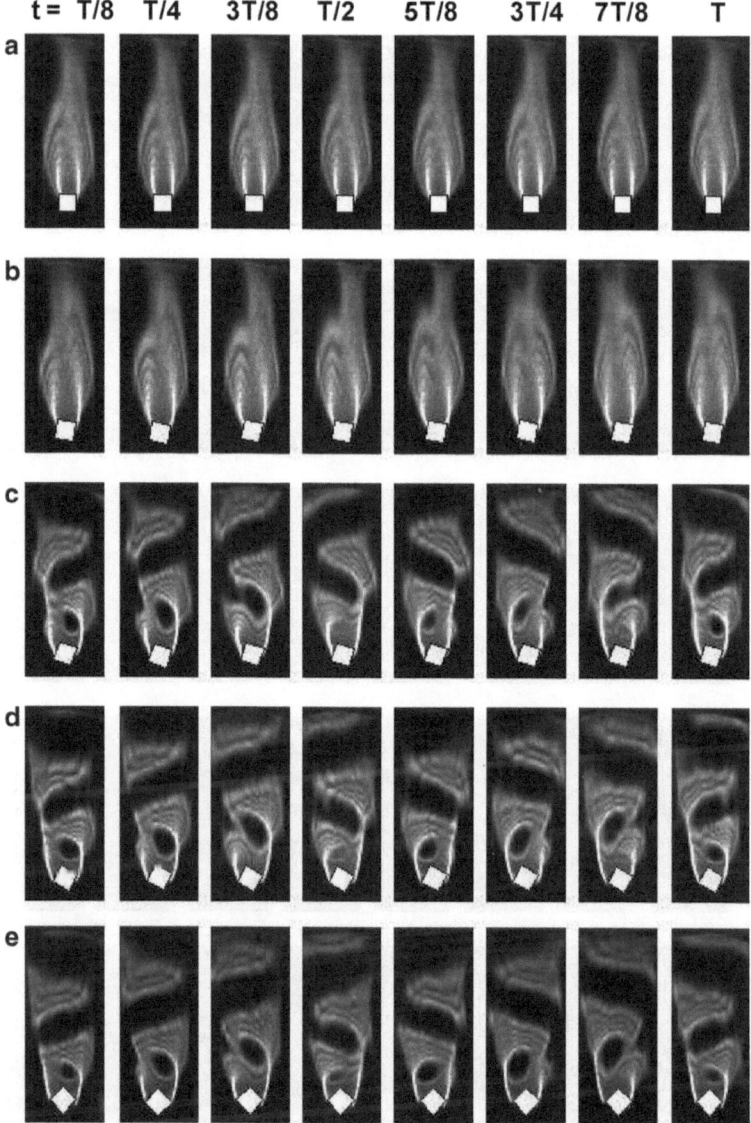

Fig. 2.13 Instantaneous schlieren images separated by a time interval of one-eighth of the time period of vortex shedding at $Re = 100$ and $Ri = 0.182$ for incidence angles of (**a**) 0, (**b**) 10, (**c**) 20, (**d**) 30, and (**e**) 45 °. The time period of vortex shedding is denoted as T

classified as *passive* and *active*. The former refers to modifying the geometry and texture of the surface itself. Experiments show that substantial benefits can be derived if the control strategy is tailored to the characteristics of the flow field, particularly after the fluid and the surface have had an interaction. This flow field

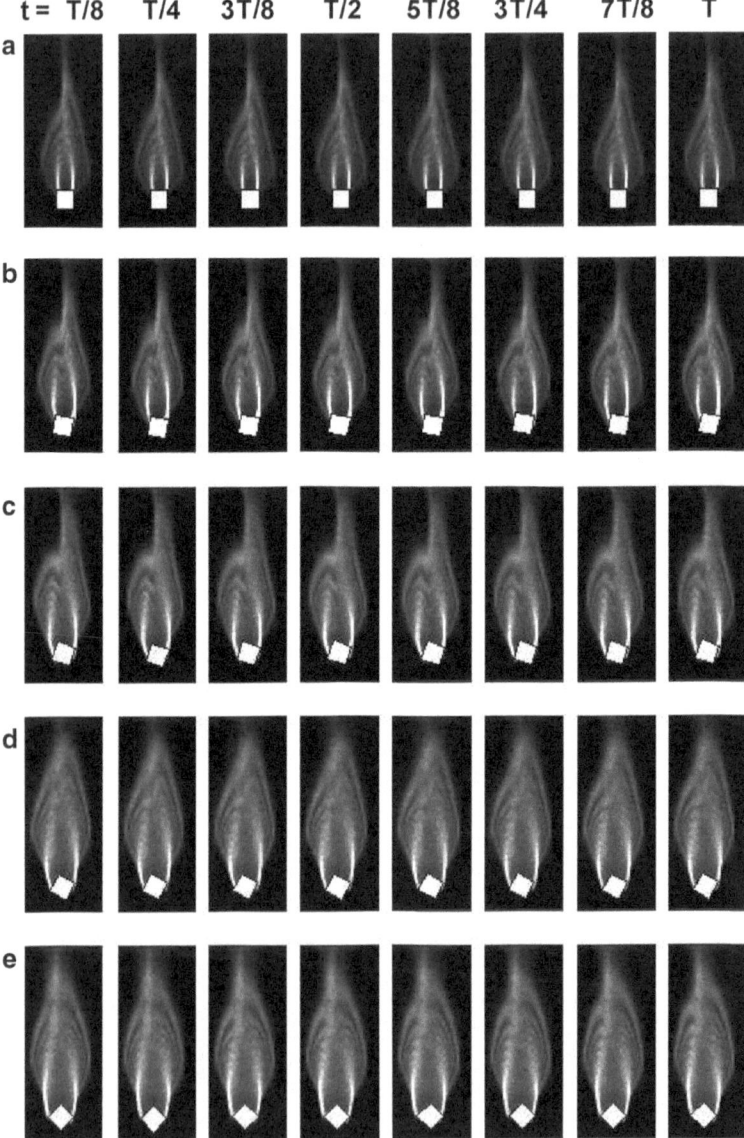

Fig. 2.14 Instantaneous schlieren images separated by a time interval of one-eighth of the time period of vortex shedding at $Re = 100$ and $Ri = 0.248$ for incidence angles of (**a**) 0, (**b**) 10, (**c**) 20, (**d**) 30, and (**e**) 45 °. The time period of vortex shedding is denoted as T

is, in most contexts, filled with three-dimensional vortices moving away from the surface. The idea is to employ an active flow control approach for modifying the vortices in the wake of the object. One approach is to oscillate the cylinder with the right amplitude, frequency, and phase [4, 8, 9, 13]. The success of the control technique is best gauged if the vortices are visualized, say, by an optical technique such as schlieren.

The effect of inline oscillations on the wake properties of square and circular cylinders is discussed in the following sections.

2.6.1 Inline Oscillations

Results obtained from schlieren imaging are presented in the following sequence: (a) instantaneous schlieren images, (b) time traces, and (c) power spectra.

Figure 2.15 show the instantaneous schlieren images as a function of Richardson number for a square cylinder excited at the fundamental frequency fe being equal to the shedding frequency fs of a stationary cylinder. The amplitude is fixed at 6% of the edge of the cylinder. The time interval between two consecutive images is equal to one-eighth of the time period of cylinder oscillation. The first image in the sequence is at the lowest position of the cylinder. Figure 2.15a, b show the instantaneous schlieren images for an oscillating cylinder at a cylinder surface temperature of 46 °C ($Ri = 0.044$). In Fig. 2.15a, the wake shows antisymmetric shedding, while in Fig. 2.15b, it shows symmetric shedding. These two shedding modes switch abruptly from one mode to other with each mode persisting over a finite number of cycles. Mode switching is observed in a long continuous sequence composed of many number of cycles. These two modes of shedding with mode competition were also observed by Ongoren and Rockwell [4] for a circular cylinder oscillating at the fundamental Strouhal frequency. The present schlieren visualization images confirm these patterns of vortex formation for the square cylinder. Mode switching can be thought of as arising from a competition between natural vortex shedding and symmetrical perturbation given at the Strouhal frequency.

Figure 2.15c shows the instantaneous schlieren images for an oscillating cylinder at a surface temperature of 57 °C ($Ri = 0.079$). Here, the vortex patterns remain symmetric over the entire cycle. In Fig. 2.15d for a surface temperature at 66 °C ($Ri = 0.108$), the instantaneous schlieren images show symmetric patterns similar to that at surface temperature of 57 °C ($Ri = 0.079$). Hence, at higher Richardson numbers, the inline perturbation transforms the alternate vortex-shedding process into symmetric vortex structures. Figure 2.15e shows instantaneous schlieren images for an oscillating cylinder at a surface temperature of 76 °C ($Ri = 0.138$). The vortices are completely symmetrical in shape as the structures from both sides start growing at the same instant, travel simultaneously, and shed at the same instant. For a stationary cylinder at the given Richardson number, Fig. 2.6e, the shear layers were seen to have merged into a single steady plume, and vortex shedding had disappeared. Thus, heat transfer with inline oscillation from the cylinder to the ambient fluid is greater as compared to the stationary cylinder because a steady plume is transformed to time-dependent symmetric vortex shedding.

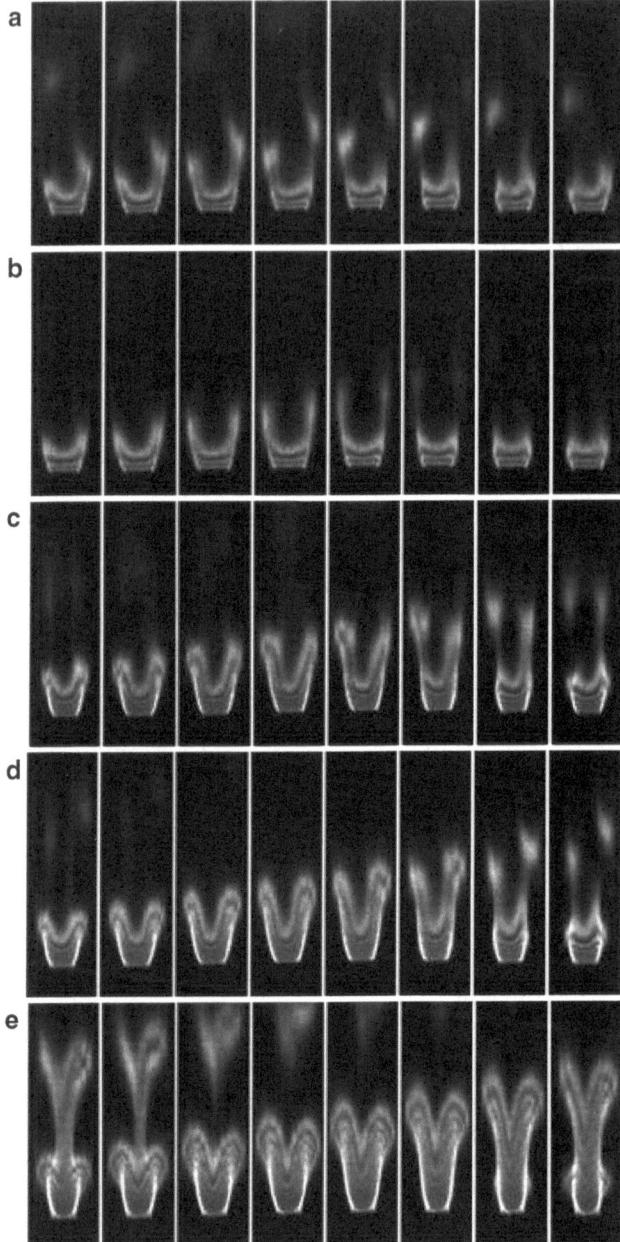

Fig. 2.15 Instantaneous schlieren images (**a**)–(**e**) for square cylinder oscillating at the fundamental Strouhal frequency separated by a time interval of one-eighth of the time period of cylinder oscillation for Richardson number; (**a**) $Ri = 0.044$ (antisymmetric mode), (**b**) $Ri = 0.044$ (symmetric mode), (**c**) $Ri = 0.079$, (**d**) $Ri = 0.108$, and (**e**) $Ri = 0.138$

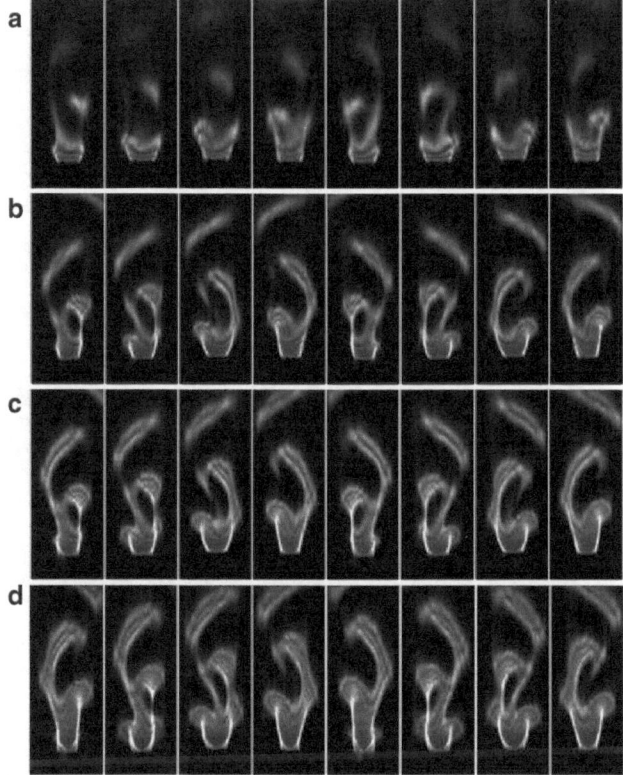

Fig. 2.16 Instantaneous schlieren images (**a**)–(**d**) for a square cylinder oscillating at the first harmonic of the vortex-shedding frequency of an unheated cylinder; images separated by a time interval of one-fourth of the time period of cylinder oscillation for (**a**) $Ri = 0.044$, (**b**) $Ri = 0.079$, (**c**) $Ri = 0.108$, and (**d**) $Ri = 0.138$

Figure 2.16 shows instantaneous schlieren images as a function of Richardson number for the cylinder oscillating at the first harmonic frequency ($fe/fs = 2$) with an amplitude of oscillation fixed at 6% of the edge of the cylinder. The images are separated by a time interval of one-fourth the time period of cylinder oscillation. The sequence of images show alternate vortex shedding for all Richardson numbers considered. The size, shape, and period of vortex structures differ from that of a stationary cylinder. For $Ri = 0.138$, vortex shedding is suppressed for the stationary cylinder (Fig. 2.6e). With harmonic oscillations, the wake is transformed from a steady plume to alternate vortex shedding (Fig. 2.16d). In Fig. 2.16, the first four images show one oscillation cycle and the next four, the next cycle of cylinder oscillation with first and fifth images at the maximum downward position of the cylinder. Careful examination of the images shows that two vortices are shed during each cycle of cylinder oscillation. The near wake is arranged into alternating pairs of vortices from either sides with each pair consisting of two vortices of opposite sign.

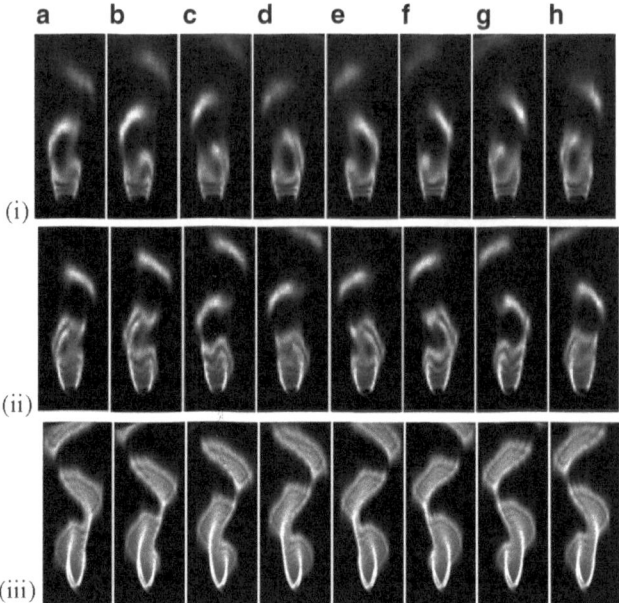

Fig. 2.17 Instantaneous schlieren images (**a**)–(**h**) at various phases of the cylinder oscillation starting from its lowest position. The cylinder frequency $fe/fs = 1$ and Richardson numbers are (i) $Ri = 0.038$, (ii) $Ri = 0.078$, and (iii) $Ri = 0.145$

Figure 2.16 shows images of two consecutive cycles with two vortices shed from opposite sides of shear layers during each cycle of cylinder oscillation. A mirror image-like appearance is to be seen.

Figure 2.17 shows the sequence of instantaneous schlieren images as a function of Richardson number for a circular cylinder oscillating transversely at the fundamental frequency ($fe/fs = 1$). Here, the excitation frequency is equal to the vortex-shedding frequency of a stationary cylinder. The time interval between two consecutive images is one-eighth of the time period of oscillation. The first image in the sequence is at the extreme leftward position of the cylinder. Images in Fig. 2.17a–h show an intense activity in the vicinity of the cylinder as compared to a stationary cylinder, and the vortices from opposite shear layers are inclined toward the wake centreline.

2.6.2 Time Traces and Power Spectra

The time traces of the light intensity fluctuations in schlieren images is presented here as a function of Richardson number. The cylinder is oscillated at the fundamental frequency (Fig. 2.18) and its first harmonic (Fig. 2.19) with an amplitude

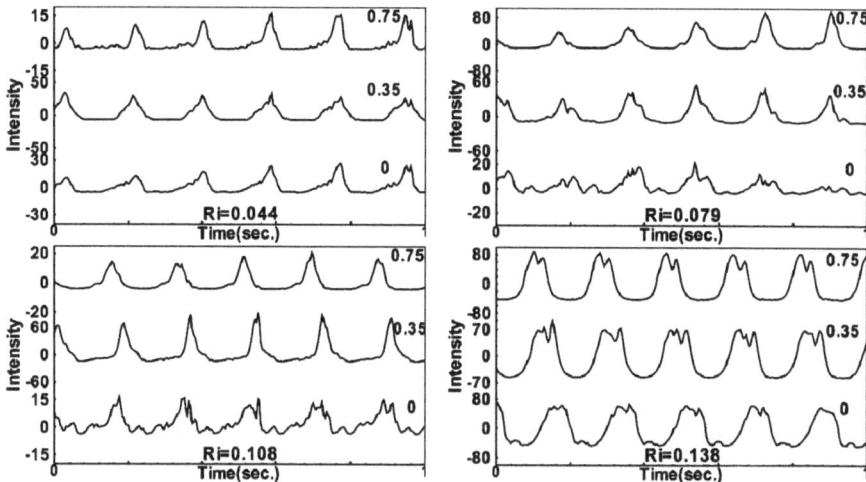

Fig. 2.18 Instantaneous light intensity versus time plotted at $y/d = 0$, 0.35, and 0.75 and various Richardson numbers for a cylinder oscillating at the fundamental frequency; $x/d = 3.5$

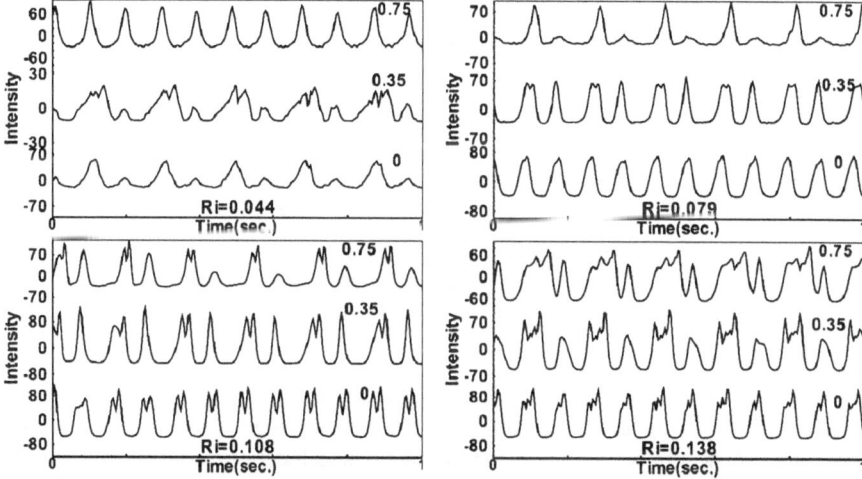

Fig. 2.19 Instantaneous light intensity versus time plotted at $y/d = 0$, 0.35, and 0.75 and various Richardson numbers for a cylinder oscillating at its first harmonic frequency; $x/d = 3.5$

equal to 6% of the edge of cylinder. The total time duration is one second, data points being drawn from 250 successive schlieren images. In Fig. 2.18, $Ri = 0.044$, and regular variation of intensity with respect to time is observed in the shear layer ($y/d = 0.35$ and 0.75). Along the midplane ($y/d = 0$), higher harmonics are superimposed over the regular intensity variation. Intensity changes with time at $Ri = 0.079$ show the variation with greater clarity as compared to $Ri = 0.044$ for all

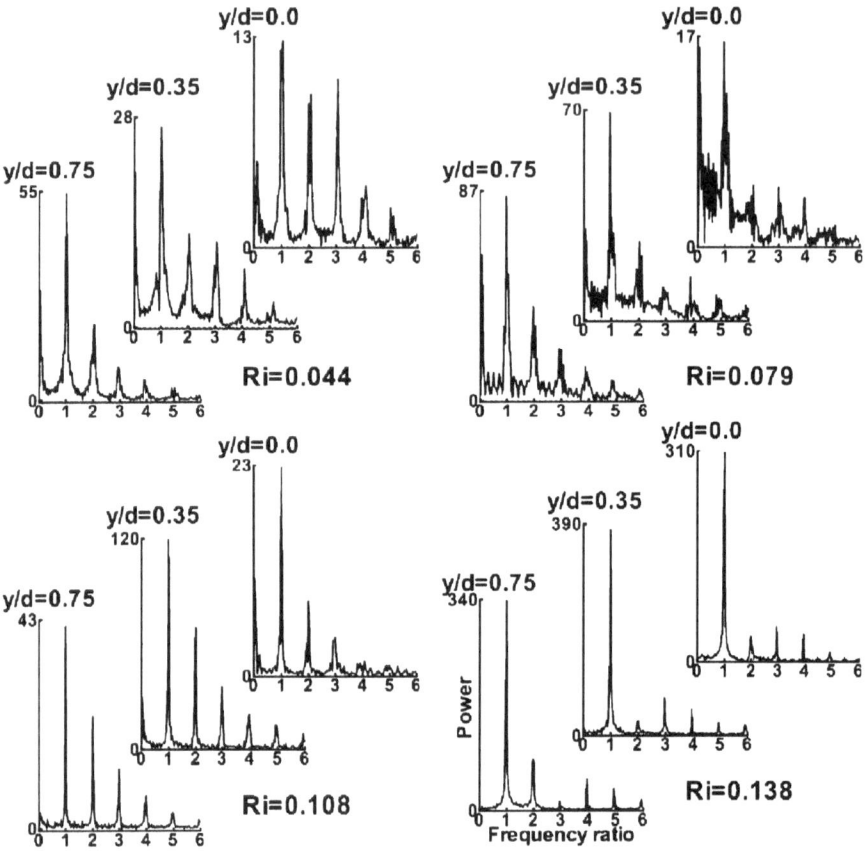

Fig. 2.20 Power spectra of light intensity at transverse locations $y/d = 0$, 0.35, and 0.75 for a square cylinder oscillating at the fundamental frequency; streamwise location considered is $x/d = 3.5$; effect of Richardson number is studied

transverse locations. Regular periodicity in the intensity variation is also observed at $Ri = 0.108$. At higher temperatures, $Ri = 0.138$, the regular large-scale periodic structures show similar variation for the transverse locations considered. This is because of symmetric structures observed in visualization images in Fig. 2.15e. When the cylinder is oscillated at the first harmonic, Fig. 2.19, $Ri = 0.044$, the variation of intensity with time is quite regular for $Ri = 0.044$ ($y/d = 0$ and 0.75). Higher harmonics are present at the location $y/d = 0.35$ due to the presence of two vortices shed within a cycle. At $Ri = 0.079$ and 0.108, intensity variations with time are regular at $y/d = 0.35$ and $y/d = 0.75$ and are related to the clear signal generated by two large vortices within one cycle. At the centerline, only one of these vortices appears, and light intensity has a near-periodic variation. This is also true for the centerline time traces at $Ri = 0.138$, while harmonics are present at $y/d = 0.35$ and $y/d = 0.75$ from two vortices shed in one cycle.

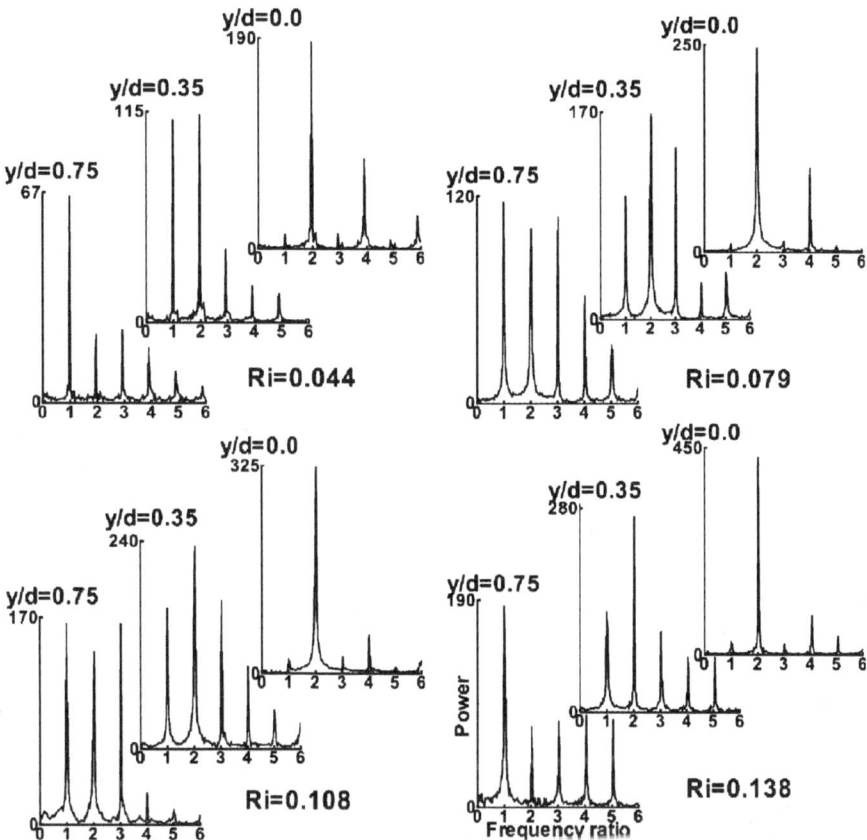

Fig. 2.21 Power spectra of light intensity at transverse locations $y/d = 0$, 0.35, and 0.75 for a square cylinder oscillating at the first harmonic frequency; streamwise location considered is $x/d = 3.5$; effect of Richardson number is studied

Power spectra of the light intensity fluctuations are shown in Figs. 2.20 and 2.21. Spectra were calculated using 1,250 images acquired in 5 s using the FFT algorithm. Power spectra can show appearance of different structures in the form of multiple peaks and can be useful in understanding the mechanisms responsible for changes in vortex shedding with an increase in Richardson number and the oscillation parameters—amplitude and frequency.

Figure 2.20 shows several spectral peaks at the three transverse locations considered. The cylinder is oscillated at the fundamental frequency, $(fe/fs = 1)$ with $Ri = 0.044$. The abrupt switching over of symmetric and antisymmetric modes can be related to the joint appearance of peaks at the fundamental frequency and its harmonics. At $Ri = 0.138$, the peak is seen at the fundamental frequency and there is complete absence of harmonics at all transverse locations. This is because the wake is filled with symmetric but time-dependent vortex structures.

In Fig. 2.21 (first harmonic, $Ri = 0.044$), the peak is seen at twice the Strouhal frequency at the cylinder centerline ($y/d = 0$). The peak at $y/d = 0.75$ is at the Strouhal frequency due to the signature of a single vortex shed in two cycles of cylinder oscillation. At $y/d = 0.35$, two comparable peaks appear at the Strouhal frequency and the first harmonic because of two vortices in one cycle. At $Ri = 0.079$ and 0.108, the peaks appear at the Strouhal frequency and its harmonic at $y/d = 0.75$ and 0.35 due to a pair of larger vortices appearing in one cycle as compared to a lower Richardson number.

2.7 Summary

Wakes of heated cylinders that are square and circular in cross-section have been imaged using the schlieren technique. The effect of cylinder orientation and oscillations on the wake have been studied. The schlieren patterns are seen to be distinct for each configuration. Results presented in this chapter show that schlieren imaging is a powerful tool for understanding the spatio-temporal features of bluff body wakes.

References

1. H.M. Badr, Laminar combined convection from a horizontal cylinder-parallel and contra flow regimes, Int J Heat Mass Transfer, Vol. 27(1), pp. 15–27, 1984.
2. A. Ben-Yaker and R.K. Hanson, Ultra-fast-framing schlieren system for studies of the time evolution of jets in supersonic cross flows, Expt. in Fluids, Vol. 32, pp. 652–666, 2002.
3. J. M. Chen and C. H. Liu, Vortex shedding and surface pressure on a square cylinder at incidence to a uniform air stream, Int. J. Heat Fluid Flow, Vol. 20, p. 592, 1999.
4. O. Cetiner and D. Rockwell, Streamwise oscillations of a cylinder in a steady current, Part 1. locked-on states of vortex formation and loading, J. Fluid Mech., Vol.427, pp.1–28, 2001. Also, A. Ongoren and D. Rockwell, Flow structure from an oscillating cylinder, part 2. Mode competition in the near wake, J. Fluid Mech., Vol.191, pp.225–245, 1988.
5. K.S. Chang and J.Y. Sa, The effect of buoyancy on the vortex shedding in the near wake of a circular cylinder, J Fluid Mech., Vol. 220, pp. 253–266, 1990.
6. F. Dumouchel, J.C. Lecordier and P. Paranthoen, The effective Reynolds number of a heated cylinder, Int J Heat Mass Transfer, Vol. 41(12), pp. 1787–1794, 1998.
7. S. Dutta, K. Muralidhar, and P. K. Panigrahi, Influence of the orientation of a square cylinder on the wake properties, Expt. Fluids, Vol. 34, p. 16, 2003.
8. C. Gau, S.X. Wu, and H.S. Su, Synchronization of vortex shedding and heat transfer enhancement over a heated cylinder oscillating with small amplitude in streamwise direction, J. Heat Transfer Trans. ASME, Vol.123, pp.1139–1148, 2001.
9. R. Govardhan and C. H. K. Williamson, Mean and fluctuating velocity fields in the wake of a freely vibrating cylinder, J. Fluids Struct., Vol. 15, p. 489, 2001.
10. O. M. Griffin, A note on bluff body vortex formation, J. Fluid Mech., Vol. 284, p. 217, 1995.
11. K. Hatanka and M. Kawahara,, A numerical study of vortex shedding around a heated-cooled cylinder by three-step Taylor-Galerkin Method, Int. J. Numerical For Methods in Fluids, Vol. 21, pp. 857–867, 1995.

12. D. R. Jonassen, G. S. Settles, and M. D. Tronosky, Schlieren PIV for turbulent flows, Opt. Lasers Engg., Vol. 44, p. 190, 2006.
13. W. Kim and J. Y. Yoo, Dynamics of vortex lock-on in a perturbed cylinder wake, Phys.. Fluids, Vol. 18, 074103(1–22), 2006.
14. C. W. Knisely, Strouhal numbers of rectangular cylinders at incidence: A review and new data, J. Fluids Struct., Vol. 4, p. 371, 1990.
15. E. Konstantinidis, S. Balabani, M. Yianneskis, The effect of flow perturbations on the near wake characteristics of a circular cylinder. J. Fluids Struct., Vol. 18, pp. 367–386, 2003.
16. J.C. Lecordier, L.W.B. Browne, S.L. Masson, F. Dumouchel and P. Paranthoen, Control of vortex shedding by thermal effect at low Reynolds numbers, Expt. Th. Fluid Sc., Vol. 21, pp. 227–237, 2000.
17. C. Lin and S.C. Hsieh, Convection velocity of vortex structures in the near wake of a circular cylinder, ASCE J. Engg. Mech., Vol. 129(10), pp. 1108–1118, 2003.
18. S.C. Luo, Y.T. Chew and Y.T. Ng, Characteristics of square cylinder wake transition flow, Phys.. Fluids, Vol. 15(9), pp. 2549–2559, 2003.
19. W.J.P.M. Maas, C.C.M. Rindt CCM and A.A. van Steenhoven, The influence of heat on the 3D-transition of the von Karman vortex street, Int J Heat Mass Transfer, Vol. 46, pp. 3069–3081, 2003.
20. J.H. Merkin, Mixed convection from a horizontal circular cylinder, Int J Heat Mass Transfer, Vol. 20, pp. 73–77, 1977.
21. N. Michaux-Leblond and M. Belorgey, Near wake behavior of a heated circular cylinder: viscosity-buoyancy duality, Exp. Therm. Fluid Sci., Vol. 15, pp. 91–100, 1997.
22. K. Noto, H. Ishida H and R. Matsumoto, A breakdown of the Karman vortex street due to natural convection, pp. 348–352, *Flow Visualization,* Springer, Berlin, 1985.
23. B. W. van Oudheusden, F. Scarano, N. P. van Hinsberg, and D. W. Watt, Phase-resolved characterization of vortex shedding in the near wake of a square-section cylinder at incidence, Expt. Fluids, Vol. 39, p. 86, 2005.
24. B.S.V. Patnaik, P.A.A. Narayana and K.N. Seetharamu,Numerical simulation of vortex shedding past a circular cylinder under the influence of buoyancy, Int. J. of Heat and Mass Transfer, Vol. 42, pp. 3495–3507, 1999.
25. J. Robichaux, S. Balachandar and S.P. Vanka, Three dimensional Floquet instability of the wake of a square cylinder, Phys. Fluids, Vol. 11(3), pp. 560–578, 1999.
26. M. Schumm, E. Berger and P.A. Monkewitz, Self-excited oscillations in the wake of two-dimensional bluff bodies and their control. J Fluid Mech., Vol. 271, pp. 17–53, 1994.
27. S. K. Singh, P. K. Panigrahi, and K. Muralidhar, Effect of buoyancy on the wakes of circular and square cylinders: A schlieren-interferometric study, Expt. Fluids, Vol. 43, p. 101, 2007. Also, A. Kakade, S. K. Singh, P. K. Panigrahi, and K. Muralidhar, Schlieren investigation of the square cylinder wake: Joint influence of buoyancy and orientation, Physics of Fluids, Vol. 22(5), 054107, 01–18, (2010).
28. K.M. Smith and J.C. Dutton, A procedure for turbulent structure convection velocity measurements using time-correlated images, Expt. Fluids, Vol. 27, pp. 244–250, 1999.
29. A. Sohankar, C. Norberg, and L. Davidson, Low Reynolds number flow around a square cylinder at incidence: study of blockage, onset of vortex shedding and outlet boundary condition, Int. J. Numer. Methods Fluids, Vol. 26, p. 39, 1998.
30. A. Sohankar, C. Norberg and L. Davidson, Simulation of three-dimensional flow around a square cylinder at moderate Reynolds numbers, Phys. Fluids, Vol. 11(2), pp. 288–306, 1999.
31. A. Wang, Z. Travnicek and K.C. Chia, On the relationship of effective Reynolds number and Strouhal number for the laminar vortex shedding of a heated circular cylinder, Phys. Fluids Vol. 12(6), pp. 1401–1410, 2000.

Chapter 3
Convection in Superposed Fluid Layers

3.1 Introduction

Buoyancy-driven convection in differentially heated fluid layers overlaid on each
other is the subject of interest in this chapter. The problem is characterized by the
formation of a fluid–fluid interface. The nature of coupling between the two fluid
layers is determined by the interface. Experiments conducted over a limited range
of parameters with an air–silicone oil system are discussed in the sections below.
For the range of parameters studied, surface tension was found to be of secondary
importance.

3.1.1 Single-Layer Convection

Convection in differentially heated horizontal fluid layers is a problem of fundamental as well as practical importance [2, 5]. The flow pattern associated with this
configuration shows a sequence of transitions from steady laminar to unsteady flow
and ultimately to turbulence. The superposed fluid arrangement has been studied by
analytical and computational techniques as well as by experiments to understand
the physics involved in the transition phenomena. Several of the global features
observed by numerical solutions are supported by experimental observations. With
new interest in understanding nonlinear systems as well as the availability of high
performance computers, there is a revival in the simulation of unsteady three-
dimensional convection. Experimental techniques have been strengthened by the
availability of optical methods to visualize flow phenomena and computers for data
collection, processing, and analysis.

   In the simplest form, the flow configuration comprises a horizontal fluid layer
confined between a pair of parallel horizontal plates that forms a cavity. The fluid
is differentially heated by maintaining the lower surface at a higher temperature
(T_{hot}) compared to the top. This situation produces a top-heavy arrangement that is

P.K. Panigrahi and K. Muralidhar, *Imaging Heat and Mass Transfer Processes*,
SpringerBriefs in Applied Sciences and Technology 4, DOI 10.1007/978-1-4614-4791-7_3,
© Pradipta Kumar Panigrahi and Krishnamurthy Muralidhar 2013

unstable. The dimensionless quantity that characterizes buoyancy-driven flow is the Rayleigh number (Ra) defined as

$$Ra = \frac{g\beta(T_{hot} - T_{cold})h^3}{\nu\alpha}. \tag{3.1}$$

Here, g is the acceleration due to gravity, β is the volumetric expansion coefficient of the fluid, ν and α are kinematic viscosity and thermal diffusivity, respectively, and h is the cavity height. When the Rayleigh number is below a critical value, the gravitational potential is not sufficient to overcome viscous forces within the fluid layer. For Rayleigh numbers above the critical value, steady circulatory flow is established. Subsequently, flow undergoes a sequence of transitions, finally resulting in turbulent motion. Buoyancy-driven flow in differentially heated horizontal fluid layers is often called *Rayleigh–Benard* convection.

Transitions in Rayleigh–Benard convection depend on Rayleigh number, Prandtl number ($= \nu/\alpha$), and the cavity aspect ratio. Additionally, there is an effect of the geometric structure of the side walls being straight or curved. The present discussion is restricted to a rectangular cavity. For a fluid layer with an infinite aspect ratio, the first transition, namely, the onset of fluid motion, occurs at a Rayleigh number of 1,708, irrespective of the Prandtl number. The associated flow pattern is in the form of hexagonal cells. The general effect of lowering the aspect ratio is to stabilize the flow due to the presence of the side walls and thus increase the critical Rayleigh number. All subsequent transitions are Prandtl number dependent.

3.1.2 Two-Layer Convection

Two-layer convection is seen in confined, differentially heated, superposed layers of two fluids. The two layers can be a combination of a gas and a liquid or two immiscible liquids with the lighter one above. They are confined between two rigid horizontal surfaces in a cold-above hot-below arrangement. At first glance, two-layer convection can be thought of as an extension of the classical Rayleigh–Benard convection problem. However, the simple act of adding a second layer of gas or immiscible fluid to the first opens up a vast parameter space that needs thorough exploration. The complexity of two-layer convection is such that new flow states appear that cannot be found in the single-layer system [1, 4, 9, 11].

Convection in a two-layer system not only depends on Rayleigh and Prandtl numbers of each layer but also the layer height ratio. In addition, material property ratios based on thermal conductivity, thermal expansion coefficient, and kinematic viscosity play important roles. The main difference between one- and two-layer convection is, however, in the formation of an interface. With a liquid–gas combination, the interface is a free deformable surface. The interface shape will depend on the pressure difference across the fluid media and the interfacial tension. Similar trends would be seen in liquid-liquid combinations as well. Temperature gradients along the interface will generate surface tension gradients, and the flow pattern will be a resultant of surface tension gradient-driven convection along with the buoyancy-driven. The former is referred as Marangoni convection. The importance of this mode of convection depends on the dimensionless group called the Marangoni number defined as

$$Ma = \frac{\sigma_T \Delta T h}{\mu \alpha} \tag{3.2}$$

Here, σ_T is the change in interfacial tension with respect to temperature, α is thermal diffusivity, and μ is the dynamic viscosity of the fluid. In many contexts, Marangoni convection is opposed to the buoyancy-driven, and the resulting flow pattern can show multiple recirculatory cells in the fluid layers [3, 6, 12]. In two-layer convection, Rayleigh number is based on the layer height and the temperature difference between the interface and the appropriate wall. By examining the definitions of Rayleigh and Marangoni numbers, it is clear that Rayleigh–Benard convection will be prevalent when the layer depth h is large while Marangoni convection would be important in thin fluid layers. Interfacial tension gradient- and buoyancy-driven convection would be jointly present at intermediate layer heights.

An intriguing feature in two-layer convection is the coupling mechanism of the flow fields in each layer across the interface [7, 8]. Two distinct modes of coupling between the fluid phases are possible. These are *thermal* and *mechanical* coupling, respectively. In mechanical coupling, the circulation in one layer drives that in the other by the mechanism of viscosity. Thus, the two rolls are oppositely oriented, and flow along the interface is in same direction. In thermal coupling, recirculation patterns in individual layers are driven by temperature difference appropriate for each of them. The rolls in each layer have identical sense, clockwise or anti-clockwise and acquire opposite velocities across the interface. In order to realize the no-slip condition at the interface, a region of high shear must prevail at this location. Irrespective of the nature of coupling, the interface conditions of continuity of temperature and heat flux are fulfilled.

3.1.3 Applications

Two-layer convection is an interesting transport process, for reasons both theoretical and practical. Theoretically, the problem is one of nonlinear dynamics with a wide range of parameters. By the simple addition of a second convecting layer, qualitatively new phenomena appear in the fluid system. These include competition between instabilities in the two layers, motion of one pattern by another, deformation of the interface, and associated instabilities.

Buoyancy-driven convection in superposed fluid layers is of interest in a variety of applications. Some of them are the following:

1. Convection in earth's mantle
2. Atmospheric convection
3. Interfacial waves in ocean exposed to the atmosphere

4. Liquid-encapsulated crystal growth
5. Crystal growth from its aqueous solution
6. Steam generators of nuclear power plants

Experimental study of two-layer convection encounters a severe problem of high-dimensional parameter space. Thus, the flow phenomena cannot be fully explored, and portions of the parameter space remain unaccessed. At the same time, nonlinear effects (specifically at high Rayleigh numbers) not reliably predicted by theoretical analysis can be recorded with careful measurements in the laboratory. This is particularly applicable for interface deformation. Hence, experimental studies serve as a benchmark for theory in two-layer convection. In the following discussion, air–silicone oil (50 cSt) experiments are discussed. The focus is on the nature of coupling and the deformation of the interface. For the range of parameters studied, surface tension and Marangoni convection was found to be secondary importance.

3.2   Apparatus and Instrumentation

The experimental arrangement that studies buoyant convection in superposed layers is quite simple in appearance. The apparatus is formed by two horizontal surfaces of very high thermal conductivity that are maintained at distinct temperatures. However, the flow patterns realized are sensitive to the uniformity and constancy of surface temperatures, parallelism of the walls, parallelism between optical windows placed at the entrance and exit to the laser beam, and properties of the insulating surfaces. These factors determine the quality of the experiment and its suitability for comparison with theory. Extraneous factors such as building vibrations, air currents, and changes in the ambient temperature have a strong influence on the quality of the recorded data. Thus, experiments and imaging have to be conducted with due care and precaution.

The schematic drawing of the test cell used to study two-layer convection is shown in Fig. 3.1. It consists of three sections, namely, the top tank, the test cavity, and the lower tank. The test section is octagonal in plan and has a nominal diameter of 130.6 mm and a height of 50 mm. The plan view closely approximates a circular cavity. The fluid layers are confined by two copper plates of thickness 1.6 mm above and below. The octagonal cavity is essentially made of optical windows, 50 mm square and 3 mm thick, with 8 of them forming the octagon in plan. The windows are essential in the present work since they allow viewing of the thermal fields at parallel incidence and at various angles. For the octagonal geometry adopted for the experimental apparatus, view angles of 0, 45, 90, and 135 ° are possible. The optical windows are made of commercially available laser-grade fused silica. The high-quality windows permit the passage of the laser beam without refraction. The optical windows were joined from outside by a mixture of hard-setting glue and grease. Thus, there were no disturbances to the field within the cavity, and leaks were

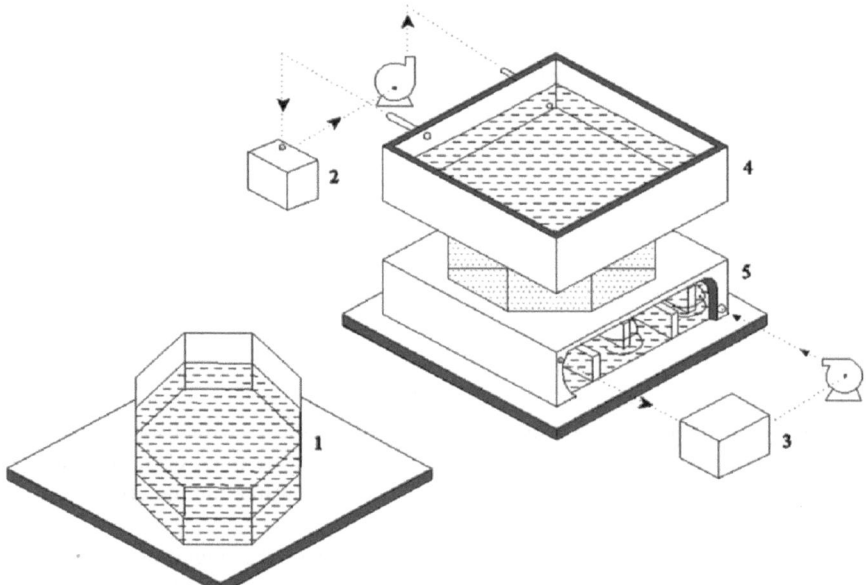

Fig. 3.1 Schematic diagram of the test cell, octagonal in plan. *1*, octagonal cavity; *2*, constant temperature bath (hot); *3*, constant temperature bath (cold); *4*, upper tank; *5*, lower tank

eliminated. Thermally active surfaces have been maintained at uniform temperatures by circulating water from constant temperature baths. On the lower tank, there is a natural tendency for the circulating water to form a free surface below the lower copper sheet. This necessitates the use of baffles suspended from the copper sheet that act as fins and maintain the temperature of the lower wall at the desired value. For the top tank, contact area between the flowing water and the copper surface is readily available. Both horizontal surfaces have been maintained at their respective temperatures to within ±0.1 K, the tolerance of temperature control obtainable with the constant temperature baths. The heat transfer areas of the two tanks have been made roughly 1.4 times the cavity surface area to reduce edge effects. To resolve the near wall intensity variation, the two copper plates have been carefully checked for flatness and surface finish. For measurements in liquids, a reference chamber is required for interferometric measurements that will compensate for refractive index changes under isothermal conditions. The reference cell utilized in the present experiments is rectangular in construction. It is placed in the compensation chamber of the interferometer and is thermally inactive. By including the reference chamber, the interferograms reveal exclusively the variations in the thermal field in the cavity. Compensation chamber is not required for schlieren and shadowgraph imaging. The components of the optical measurement system include 60 mW He–Ne laser, 150-mm-diameter optical elements, CCD camera with 512×512 resolution, 8-bit frame grabber, and image acquisition at 25 frames-s^{-1}.

Experiments have been carried out within a cavity half-filled each with air and silicone oil. The viscosity of silicone oil used in the experiments was 50 cSt (Pr = 11). Temperature differences over the range of 5–22 K were utilized across the cavity. The temperature difference across each layer was found to scale with the respective conduction resistance. Accordingly, large temperature differences were realized in air while smaller values appeared in oil. The average interface temperature at steady state was determined iteratively by requiring the average heat flux through each fluid layer to be a constant. The overall cavity temperature difference had to be kept below 5 K in interferometry experiments owing to refraction errors. This restriction did not apply to shadowgraph measurements. Marangoni numbers were calculated to be in the range of 10–100 and hence small enough to be of no significance. Rayleigh numbers obtained in the three fluid layers are around 15,000 in air and 65,000 in oil, making buoyancy the most dominant driving force for fluid convection. The discussion below is focused on interface deformation and its correlation with the local wall heat transfer rates. For a longer discussion on other fluid combinations such as oil and water and the effect of layer height, the reader is referred to [8] and [10].

Experiments commenced with the fluid layers and the bounding walls all at the ambient temperature. Water from the constant temperature baths were supplied to the two horizontal bounding surfaces, indicating a step increase in the overall applied temperature difference. Steady state patterns, if reached, were recorded at the end of 4 h. Otherwise, the dominant pattern that prevailed for the longest duration was recorded. At large temperature differences, persistent unsteadiness was to be seen. Longtime unsteadiness appearing after about 5–6 h of experimentation was of interest. Here, images were recorded as a time sequence at video rates of the CCD camera.

3.3 Unsteady Two-Layer Convection

The present discussion is concerned with unsteadiness and three dimensionality in differentially heated superposed fluid layers. The objective is to investigate interface deformation and its effect on wall heat transfer. In the absence of Marangoni convection, Rayleigh number plays a central role in determining the strength of convection. In dimensional form, deformation of the interface depends on the overall temperature difference and the individual heights of the fluid layers. When heated from below, buoyant plumes move from the lower wall toward the interface. For a small temperature difference (ΔT) across the cavity, convection is not strong enough to cause deformation of the interface. For moderate ΔT, the rising plume tends to push the interface, leading to the formation of a crest and a trough. The deformed shape of the interface does not change with time. The flow field continues to be steady, though it may be three dimensional. For a larger temperature difference, interface deformation becomes a function of both position and time.

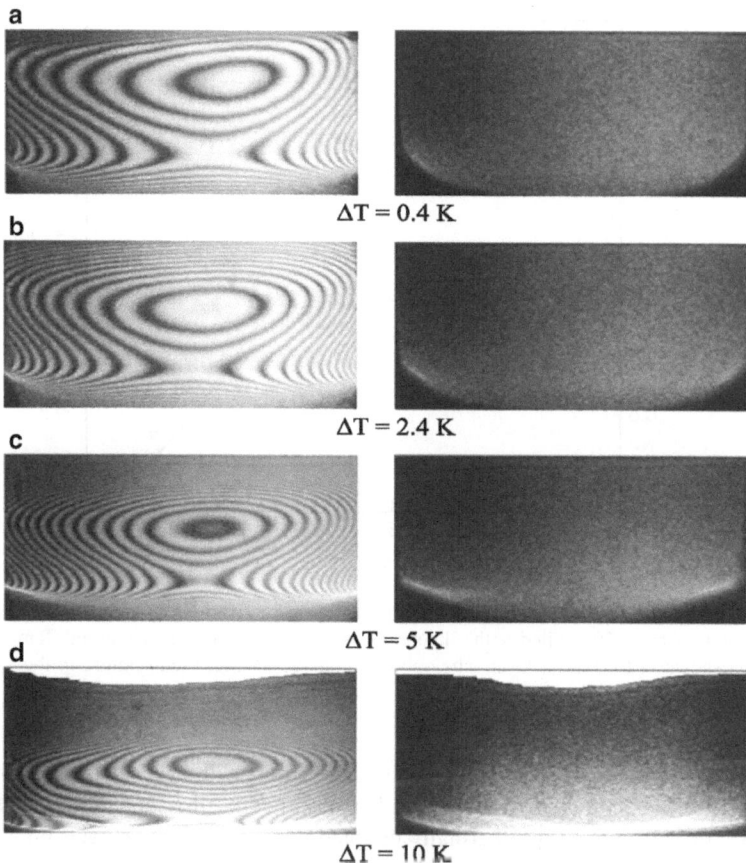

Fig. 3.2 Interferograms and shadowgraphs in silicone oil that show the effect of increasing overall temperature difference across the cavity that is 50% filled with silicone oil, the rest being air. The portion of the cavity containing air is not shown. Interface deformation is seen in (**d**)

Figure 3.2a–d shows interferograms and shadowgraph images obtained for various temperature differences across the cavity when it is half-filled with silicone oil, the rest being air. For an overall temperature difference of 10 K, the Rayleigh numbers in air and silicone oil were calculated as 12,244 and 39,425, respectively. As the overall temperature difference across the cavity increases, the temperature change across silicone oil increases, though a greater temperature drop takes place in air. Hence, there is a continuous increase in Rayleigh number for each fluid with temperature difference across the cavity. With an increase in Rayleigh number, Fig. 3.2 shows that the number of fringes increases. The fringe density near the interface also increases indicating an increase in the heat flux at the air-oil interface. The fringes are practically aligned with the interface. Thus, the interface tends to become an isotherm. For this reason, convection driven by surface tension gradients along the interface is expected to be small in the present work. The comparison

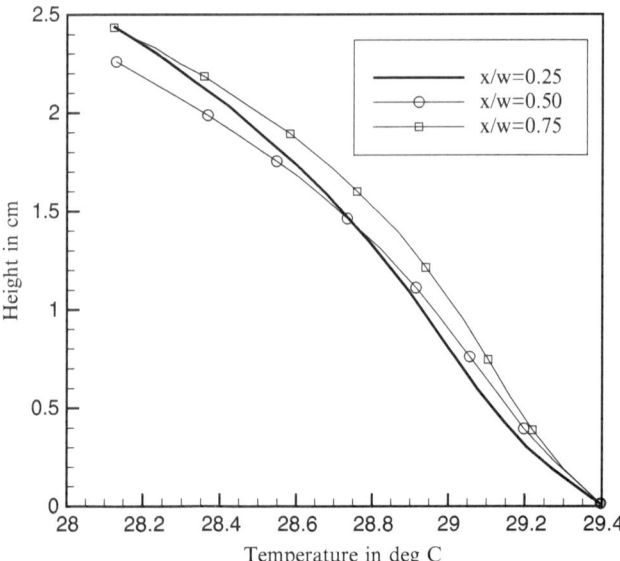

Fig. 3.3 Temperature profiles at a particular time instant along various columns of the cavity for $\Delta T = 10\,^\circ\text{C}$. Cavity 50% filled with silicone oil is considered, the rest being air. Temperature variation shown is for the oil layer. The optical technique averages temperature in the viewing direction

is against the contribution of buoyancy that scales with the cavity temperature difference. The high fringe density and, consequently, the large interfacial flux suggest that air and oil are thermally coupled, the convection in each fluid layer being driven by their respective temperature differences. The sum of the two temperature differences is equal to that applied across the cavity. For an increase in temperature difference, Fig. 3.2 shows that the center of the Ω-shaped fringe moves away from the interface and toward the lower wall. The formation of these distinctive fringes has been discussed by the authors in [8] and [10]. Also see [10] of Chap. 1.

Though not noticeable at lower temperature differences, Fig. 3.2 shows visible interface deformation at a temperature difference of 10 K.

Figure 3.3 shows the temperature variation within the oil layer at a given time instant for three different locations. Temperatures have been calculated from the shadowgraph images of Fig. 3.2d. Temperatures at the interface for the three columns are nearly equal, but a change in interface location is visible.

Figure 3.4 shows interferograms and shadowgraph images obtained at $\Delta T = 5$ K for four different view angles. The cavity is filled with 50% silicone oil, leading to a Rayleigh number of 6,106 and 20,062 in air and oil, respectively. In the interferograms, fringes very near the hot wall are lost due to refraction. The lower wall in turn appears to be deformed. The interface, however, appears to be quite flat. Owing to a high fringe density at the interface, interferometric analysis was not possible. Results based on shadowgraph evaluation are reported in the following

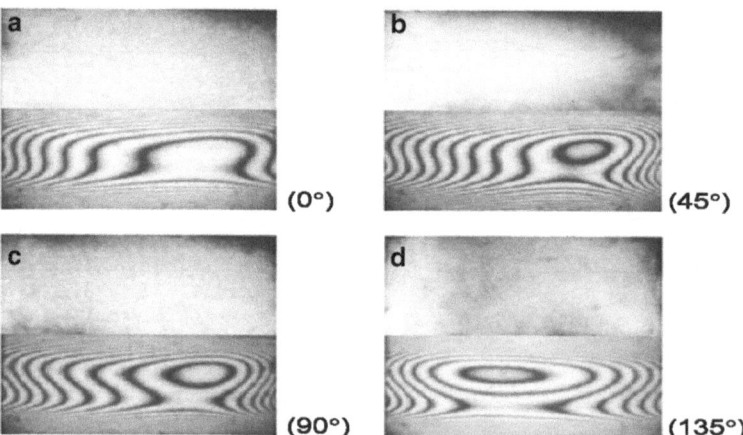

Fig. 3.4 Interferograms and shadowgraphs for an overall temperature difference of 5 K. The cavity is 50% filled with silicone oil (where fringes are visible), the rest above the oil layer being air. View angles considered are 0, 45, 90, and 135°, respectively. Clear changes in light intensity were seen only at higher temperature differences

section. The interferograms show clear differences among the four view angles. The difference can be ascertained from the location of the center of the Ω-shaped fringe. Thus, it can be concluded that the convective field in silicone oil is three dimensional.

The correlation between interface deformation and the local interfacial heat flux is examined next. The analysis is performed with the help of the instantaneous shadowgraph images. The interfacial flux is reported in dimensionless form and equivalently in terms of the local Nusselt number as a function of the coordinate in the positive x-direction. The Nusselt number variation along the lower heated wall is also included for comparison. Figure 3.5 presents the variation of Nusselt number over the interface and the lower wall along with interface deformation for ten different time instants. Figure 3.6 presents the timewise variation of Nusselt number at the interface and the heated wall at three different x-locations. The instantaneous plot with respect to the spacecoordinate and the temporal variations shows a strong association among interface flux, wall flux, and interface deformation. From Figs. 3.5 and 3.6, the following conclusions can be drawn:

1. Nusselt number at the interface has a positive correlation with interface deformation. Thus, regions of large deformation from the mean position are also regions of high interfacial heat flux. Conversely, the downward movement of the interface lowers the interfacial flux.
2. The correlation between interface deformation and Nusselt number is seen in instantaneous shadowgraph images as well as in the entire time sequence.
3. The wall Nusselt number negatively correlates with the interface deformation for a given streamwise position along the cavity.

54

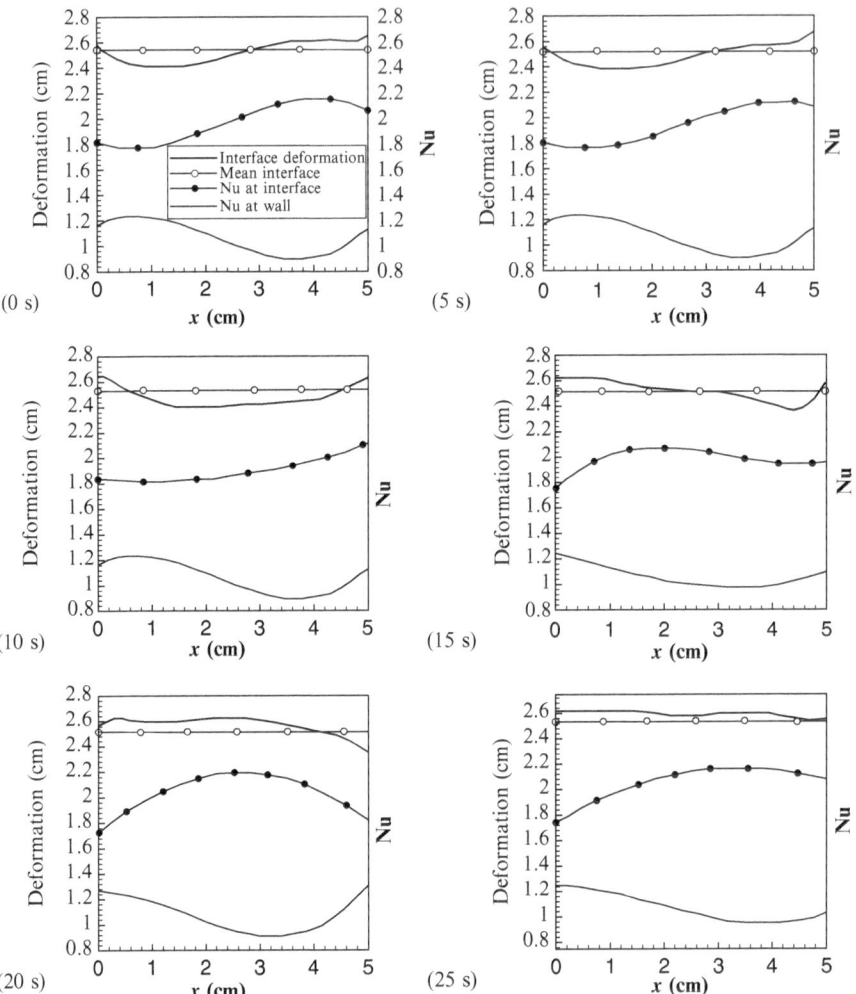

Fig. 3.5 Interface deformation and Nusselt number distribution over the interface and the lower hot wall for various time instants between 0 and 25 s with a cavity temperature difference of $\Delta T = 10\,°C$; the cavity is 50% filled with silicone oil, the rest being air

The above observations can be explained in the following manner. The vertically upward movement of the interface increases its distance from the lower wall and thus increases the conduction resistance. Consequently, a reduction in the wall heat flux is to be expected. The shadowgraph experiments reveal an opposite trend for the following reason: The crest of the interface and the point of maximum heat flux at the lower wall are both parts of a single recirculation loop created by buoyancy. The circulation pattern communicates the peak heat flux from the wall to the highest point on the interface, creating a joint maximum in displacement

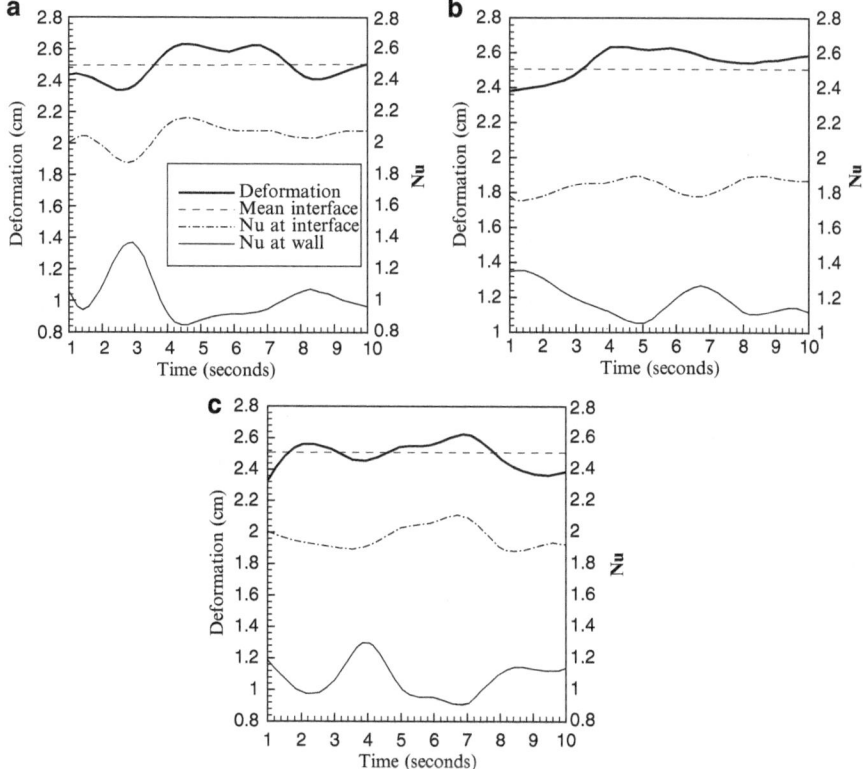

Fig. 3.6 Temporal variation of interface deformation compared with Nusselt numbers at the interface and the lower wall for three dimensionless locations scaled by the cavity height: (**a**) $x = 0.25$, (**b**) $x = 0.5$, and (**c**) $x = 0.75$. Overall temperature difference is $\Delta T = 10\,^\circ\mathrm{C}$, and the cavity is 50% filled with silicone oil, the rest being air

and Nusselt number at that location. Directly below the trough, the fluid motion is less vigorous and a minimum in the heat flux is obtained. A second factor that contributes to the positive correlation is fluid velocity. At the crest of the interface, it is expected that the x-component velocity is high, resulting in large heat fluxes across the interface. For a similar reason, the heat flux reaches a minimum below a trough in interface displacement. The correlation between heat flux and interface displacement was observed in sequences of images recorded in experiments for a given temperature difference. It was also to be seen in all the four view angles at which shadowgraph imaging was conducted. Experiments conducted with a large temperature difference across a cavity with 30% silicone oil revealed visible interface deformation. A shadowgraph analysis of the recorded images once again showed a correlation between interface movement and the interfacial flux. For a cavity with 70% oil, the interface movement was very small, and the correlation could not be ascertained.

3.3.1 Three-Dimensional Reconstruction of Nusselt Number

The Nusselt number distribution of the previous section is an average evaluated
along the viewing direction. The present section is concerned with obtaining the
local distribution of Nusselt number over the entire interface. A similar distribution
at the lower wall is also of interest. The reconstruction of Nusselt number over the
interface has been accomplished by first recording shadowgraph images along four
distinct view angles. Interpreting each view as a path integral and applying tomographic algorithms, specifically, convolution back projection (CBP), the variation
of Nusselt number over an entire surface has been obtained. Tomography can be
applied in principle only for time invariant fields. In the present work, tomography
has been applied to specific modes of convection that are in turn determined by
a proper orthogonalization procedure. The input to these calculations is a time
sequence of shadowgraph images for each view angle.

Proper orthogonal decomposition (POD) is a powerful method of data analysis
that converts a function of both time and space into a product of functions that
depend individually on space and time (also see [10] of Chap. 1). Functions of
space thus obtained are the modes contained in the data set. The decomposition is
performed in an optimal manner in such a way that the lowest modes contain most
information about the entire time sequence. The zeroth-order mode is obtained as an
average of the time sequence of shadowgraph images. In the present experiments,
the time sequence of shadowgraph images are converted into modes of various
orders for each view angle. It is assumed that modes of identical order for each
of the viewing directions uniquely correspond to one another and hence can be
used for tomography. Both POD and tomography calculations have been performed
using MATLAB. The numerical procedure has been validated against simulated data
where the field of the dependent variable was taken as a function of x, y, and time.
The errors in reconstruction were found to be less than $\pm 1\%$.

Figure 3.7 shows the zeroth-order Nusselt number distribution over the interface
and the lower (heated) wall. Figure 3.7 also shows the first and the second modes of
Nusselt number for the interface and the lower wall. In continuation to the discussion
of Sect. 3.3.1, the interfacial Nusselt number distribution is seen to correlate
positively with the interface deformation. The bright patch in the reconstructed
gray-scale image of Nusselt number indicates high values of the interfacial flux.
Figure 3.7 shows that bright regions in the central portion of the interface correspond
to dark regions at the lower wall. Darker regions at the interface also correlate with
brighter regions at the lower wall. These results are generalizations of the trend seen
in Figs. 3.5 and 3.6. The flux-deformation correlation holds for both the first and the
second modes.

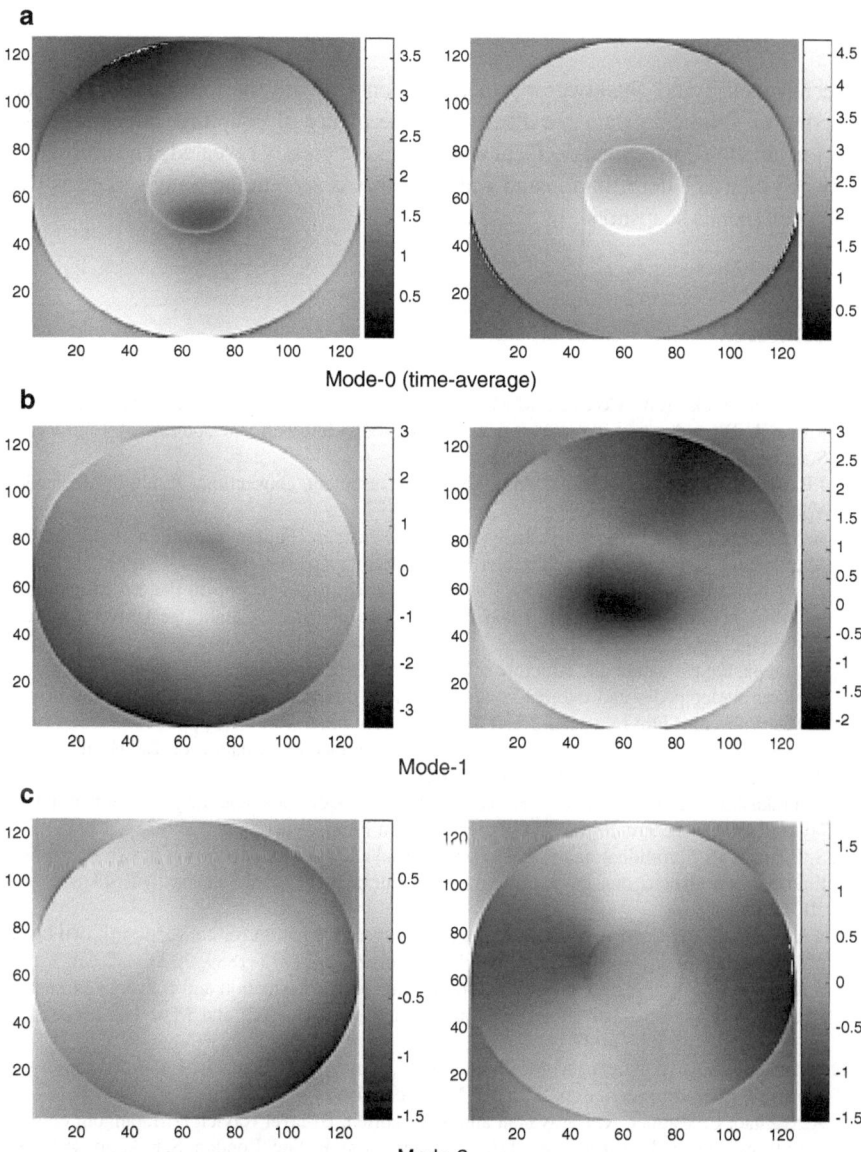

Fig. 3.7 Nusselt number distribution corresponding to Modes 0–2 at the lower wall (*left*) and the interface (*right*). Bright patches correspond to high values of Nusselt number and dark patches to low. The *dark and bright patches* in the left and right columns appear jointly and indicate a definite correlation

3.4 Summary

Interferometric and shadowgraph measurements of the thermal fields in an air–silicone oil 2-layer system are discussed. The nature of coupling, interface deformation, and the corresponding local wall fluxes are clearly revealed. The correlation between interface deformation and wall heat flux is seen in the timeaverage as well as instantaneous snapshots.

References

1. C.D. Andereck, P.W, Colovas, M.M. Degen and Y.Y. Renardy, Instabilities in two-layer Rayleigh–Benard convection: overview and outlook, International Journal of Engineering Science, Vol. 36, pp. 1451–1470, 1998.
2. J.P. Gollub and S.V. Benson, Many routes to turbulent convection, J. Fluid Mechanics, Vol. 100(3), pp. 449–470, 1980.
3. A.A. Golovin, A.A. Nepomnyashchy and L.M. Pismen, Pattern formation in large scale Marangoni convection deformable interface, Physica D, Vol. 81, pp.117–147, 1995.
4. D. Johnson, R. Narayanan and P.C. Dauby, The effect of air height on pattern formation in liquid-air bi-layer convection, in *Fluid Dynamics at Interfaces,* Cambridge University Press, Cambridge, pp. 15–30, 1999.
5. M. Lappa, On the nature and structure of possible three dimensional steady flows in closed and open parallelepipedic and cubical containers under different heating conditions and driving forces, Journal of Fluid Dynamics and Material Processing, Vol. 1(1), pp. 1–19, 2005.
6. G. Lebon, P.C. Dauby and V.C. Regnie, Role of interface deformation on Be-Ma instability, Acta Astronautica, Vol. 48(5–12), pp. 617–627, 2001.
7. A. Prakash,K. Yasuda, F. Otsubo, K. Kuwahara and T. Doi, Flow coupling mechanism in two-layer Rayleigh–Benard convection, Experiments in Fluids, Vol. 23, pp. 252–261, 1997.
8. S. Punjabi, K. Muralidhar and P.K. Panigrahi, Buoyancy-driven convection in two superposed fluid layers in an octagonal cavity, International Journal of Thermal Sciences, Vol. 43, 849–864, 2004.
9. Y.Y. Renardy and C.G. Stoltz, Time dependent pattern formation for convection in two layers of immiscible fluids, Inter. J. Multiphase Flow, Vol. 26, pp. 1875–1889, 2000.
10. S.K. Sahu, K. Muralidhar and P.K. Panigrahi, Interfacial deformation and convective transport in differentially heated air-oil layers, Journal of Fluid Dynamics and Materials Processing, Vol. 3(3), pp. 265–286, 2007.
11. W. Shyy, H.S. Udayakumar, M.M. Rao and R.W. Smith, *Computational Fluid Dynamics with Moving Boundaries,* Taylor and Francis, New York, 1996.
12. A.X. Zhao, C. Wagner, R. Narayanan and R. Friedrich, Bi-layer Rayleigh-Marangoni convection: transitions in flow structures at the interface, Proc. R. Soc. London, Ser. A, pp. 451–487, 1995.

Chapter 4
Transport Phenomena in Crystal Growth

4.1 Introduction

The importance of crystals in modern technology can hardly be over-emphasized. They are the workhorses that drive opto-electronic devices of the present photonics era. The quality of crystals needed for optical applications has to be very high. Significant progress in crystal growth process has resulted in giant strides in several fields employing devices based on crystals. Crystal growth is seen as a multidisciplinary field involving subjects from the basic sciences as well as engineering. Competence in solid-state physics, chemistry, thermodynamics, fluid mechanics, crystallography, mathematical modeling, optical engineering, and electronics and instrumentation is jointly required. This chapter describes convection patterns formed during growth of a crystal from an aqueous supersaturated solution. The role rotation of the crystal during growth is examined.

4.2 Why Study Crystal Growth?

Crystal growth is a first-order phase transition and is associated with the liberation of latent heat. An efficient transfer of heat liberated from the solid–fluid interface region to the bulk of the solution is of major concern during crystal growth. Additionally, the processes of mass transfer and the convection modes adopted to achieve the desired mass transfer rates present a difficult optimization problem. These transport processes significantly affect the compositional homogeneity, surface microstructure and the growth rate of the crystal. Thus, crystal growth is governed by the principles of physicochemical hydrodynamics. The technology of crystal growth involves designing and developing the necessary apparatus and instrumentation for achieving controlled phase transition that enables the growth of single-crystalline solids [18].

P.K. Panigrahi and K. Muralidhar, *Imaging Heat and Mass Transfer Processes*,   59
SpringerBriefs in Applied Sciences and Technology 4, DOI 10.1007/978-1-4614-4791-7_4,
© Pradipta Kumar Panigrahi and Krishnamurthy Muralidhar 2013

To ensure growth of large high-quality crystals, it is important to understand the related convection and transport phenomena [17, 21, 25, 26]. The present chapter reports the use of schlieren and shadowgraphy techniques in this context for the growth of potassium dihydrogen phosphate (KDP) and lysozyme protein crystals. The experimental apparatus and the growth procedure are described. The influence of control parameters is brought out from the visualization images.

4.3 Crystal Growth from Solution

Crystals for optical applications are grown mainly by three different techniques, namely, solution growth, melt growth, and flux growth, depending upon the material properties of the crystal to be grown. Crystal growth from a solution is quite a complex phenomenon because of several parameters simultaneously affecting the process. When the process to be imaged uses visible radiation, the medium under study has to be transparent at the chosen wavelength of radiation (632.8 nm for the He-Ne laser used in the present experiments). Among the various methods available, crystal growth from an aqueous solution satisfies this requirement. It is also easy to experiment with because of the near-ambient working temperatures involved. For these reasons, this chapter is primarily focused on crystal growth from an aqueous solution.

A crystal growing from solution creates thermal and concentration gradients in its surroundings by releasing the heat of crystallization and removing the solute near the growth surface. The resultant temperature and concentration gradients affect the perfection and stability of the crystal grown. The change of solution density with temperature ($d\rho/dT$) is negative, and the change of solution density with solute concentration ($d\rho/dC$) is positive. Depleted of salt, the solution adjacent to the crystal is lighter than that elsewhere. Therefore, crystal growth in the earth's gravitational field is accompanied by a rising buoyant convection current which envelops the crystal, is often oscillatory and unstable, and drastically modifies the concentration gradient along the growth interface. Thus, the growth history and defect structure of the crystal is a function of the time-dependent spatial distribution of the convection patterns and of the temperature and concentration profiles in the surrounding solution. To understand the growth of large defect-free crystals, it is helpful to map the spatial distribution of the solution properties during growth.

Techniques for visualizing the temperature and concentration fields in an aqueous solution have often been used to monitor crystal growth. One method for determining the saturation temperature of a solution or establishing the initiation of growth is to use shadowgraph or schlieren imaging and examine whether the convection plume rises or descends from a seed crystal. The former corresponds to solute deposition over the crystal, while the latter occurs during dissolution of the crystal into the solution. The formation of fluid inclusions is related to the convection pattern around a crystal. A necessary first step in growing large inclusion-free crystals is the optimization of the growth process by visualizing the convection patterns [23]. Since weak transient convection can have a significant effect on crystal perfection, the fluid motion must be continuously monitored. Mapping the spatial variation of salt concentration and temperature around the crystal provides information about the mechanisms of growth, the onset of instabilities, and the relation between growth conditions and crystal quality [3,5,6,17,21]. This approach is also enhanced by simultaneous measurement of crystal growth rate and mapping of growth steps and irregularities, such as hillocks, on the growing faces [8, 10, 13].

4.3.1 Phenomenology of Crystal Growth from Solution

The rate of crystallization is determined by the slower of the two stages: (1) supply of crystallization material from the solution to the crystallization surface and (2) incorporation of this material into the crystal structure, that is, growth of the crystal. If the first stage is the limiting step, the growth is said to proceed in the diffusion regime. If the second stage is the limiting step, growth is in the kinetic regime. Diffusion regime conditions are prevalent during buoyancy-driven growth. In practice, forced flow conditions may also be adopted by stirring the solution. These conditions bring the process into the kinetic regime of growth. Here, the solute is forced to move from the bulk solution toward the crystal surface. There is an increased probability of occurrence of turbulent flow. Under such conditions, the growth rate is high. Mass transfer rates and the fluid flow conditions have to be optimized so as to avoid deleterious effects such as morphological instability, spurious nucleation, and incorporation of inclusions.

The principal effect of gravity on crystal growth is to introduce several forms of buoyant convection. For all solutions of interest, density increases with solute concentration. Therefore, the solute-depleted solution around the crystal tends to rise and converge in a plume above the crystal, while the undepleted solution rises from beneath or around the crystal as shown in Fig. 4.1. The heat of crystallization also contributes to convection but to a smaller degree. A second mode of buoyant convection occurs when the solution is cooled to maintain supersaturation, since temperature gradients are also density gradients. Other modes of buoyant convection are caused by growth-stabilizing techniques such as stirring and cooling the crystal to establish a stabilizing temperature gradient. These counteract the solutal density gradient and eliminate the gradient in supersaturation. Finally, if spurious nuclei occur in a gravitational field and are not suitably removed, they can descend to the bottom of the chamber and act as convection pumps. Various modes of convection interact with each other in a complex time-dependent manner. Even in the absence of other sources of convection, flow arising from solute depletion is spatially irregular because of boundary-layer discontinuities at crystal edges and corners. It is subject to periodic oscillations and unstable responses to small transient accelerations. The influence of convection on crystal growth is seen in the following contexts.

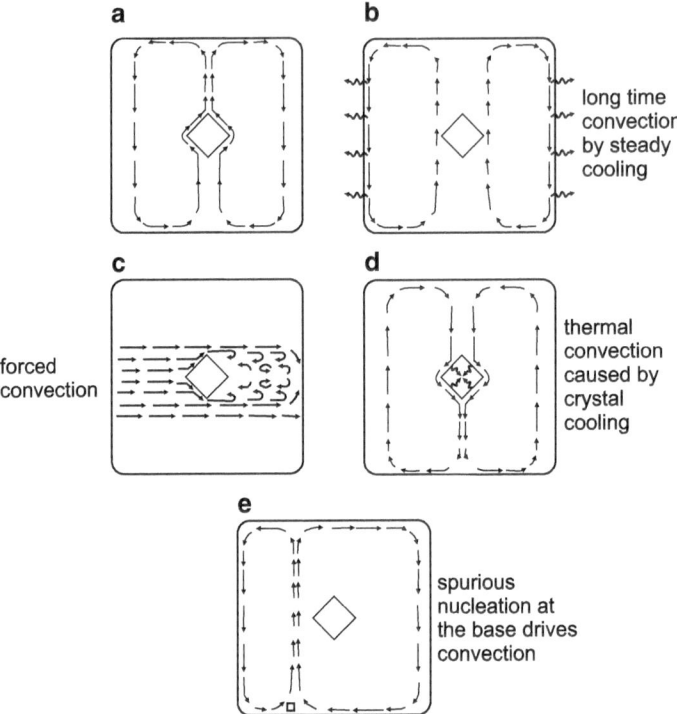

Fig. 4.1 (**a**) Solute-depleted solution around the crystal tends to rise and converge in a plume above the crystal; (**b**) second mode of buoyant convection arises when the solution is cooled to maintain supersaturation; (**c**) other modes of convection are caused by growth-stabilizing techniques such as stirring; (**d**) cooling the crystal to establish a stabilizing thermal gradient leads to convection though with a reversal in the direction of circulation; (**e**) spurious nucleation sites act as convection pumps; adapted from [23]

1. *Growth rate*: Convection greatly increases the rate of solute transport to the growth interface by replacing the diffusion zone with a much narrower boundary layer. As a consequence, the crystal growth rate increases by several orders of magnitude. Across the boundary layer, concentration increases from the saturation value toward supersaturation, as prevailing in the bulk of the solution. In practical terms, a thin boundary layer is a great advantage and is one of the reasons why forced convection is often used. Buoyant convection, on the other hand, creates differences in solute concentration above and the below the crystal and, in general, creates greater inhomogeneities in the solutal mass fluxes.

2. *Inclusions*: The relationship between convection and the formation of fluid inclusions is quite complex. In the absence of convection, fluid inclusions can form either because of interface instability or by the convergence or intersection of layers growing inward from edges and corners. Under laminar flow conditions, buoyant or forced, inclusions can form if mass transfer rates are slow enough to allow solute depletion in the boundary layers. This is another reason for the

Fig. 4.2 Fluid inclusions tend to occur as a result of a stagnation zone, on the upper portions of the side faces, and at the boundaries of convection cells; adapted from [23]

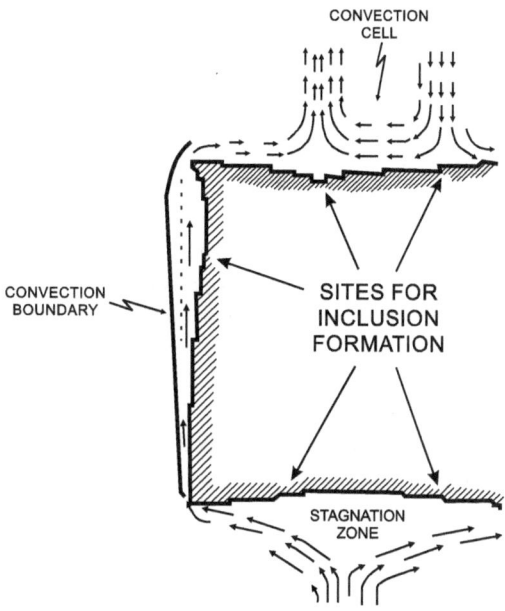

use of forced convection during the growth of large crystals. For moderately strong buoyant convection, inclusions tend to occur (a) in the stagnation zone such as the center of the lower face, (b) in the solute-depleted upper portions of the side faces, and (c) in the boundaries of convection cells on the top face (Fig. 4.2). Occasional instabilities caused by accelerational transients tend to reduce formation of fluid inclusions, perhaps by sweeping off bubbles or pockets of stagnant solution. In forced convection, it has been shown that inclusions can occur in regions where stagnant eddies are likely to occur. In summary, inclusions are intricately connected with fluid convection.

3. *Dislocations*: Dislocations are formed by mechanisms similar to those postulated for fluid inclusions.

4. *Impurity banding*: During the melt growth of semiconductor and oxide crystals, convective oscillations from temperature instabilities result in fluctuation of the doping concentration that in turn appears as *bands*. When sodium chloride crystals are grown from brine, impurity banding can be correlated to the diurnal temperature cycles driven by the convective field.

5. *Spurious nucleation*: The proliferation of spurious nuclei over a growing crystal is due to the impingement of a jet of solution on it. Such jets may arise in during circulatory motion of the solution in the crystal growth chamber.

The above discussion shows that fluid convection influences the growth rate as well as the quality of the growing crystal. Therefore, convection patterns during crystal growth are of interest. As a case study, the growth of a KDP crystal is considered here because of the availability of its thermodynamic and optical data in solution. Growth of protein crystals is subsequently presented.

4.4 Crystal Growth Apparatus

A double-walled growth cell of glass has been fabricated for performing crystal growth experiments. It has a cylindrical geometry with the cavity between the two concentric cylinders filled with thermostated water. The temperature of the KDP solution is kept within $\pm 0.01°C$ of the desired value. The PID-based programmable temperature controller (*Eurotherm*) is employed for this purpose. In order to facilitate unhindered passage of the laser beam through the growth cell, quartz optical windows are provided on opposite sides of the cell. Two ports are provided for inserting a temperature sensor and a crystal seed into the growth chamber. Figure 4.3 shows the schematic drawing of the cylindrical growth cell. The crystallizer of the present work is essentially a scaled model of the growth system used for growing large-sized KDP crystals. The crystal growth experiments were performed at the limits of free and forced convection. In order to achieve forced convection conditions, the crystal was rotated at various speeds using a stepper motor. Wobbling of the crystal growth apparatus during rotation is not desirable as it results in irregular flow in the growth cell. Also, any deviation of crystal rotation from the axis of the growth cell is detrimental to the quality of the projection data required for tomographic reconstruction. For these reasons, a specially designed apparatus has been fabricated for the forced convection experiments. A stainless steel tube of 10-mm wall thickness and 120-mm inner diameter is precisely machined at both ends to achieve the desired parallelism. A steel plate of 140-mm diameter is joined to the pipe section by screws with the stepper motor fitted above. A tool-holding

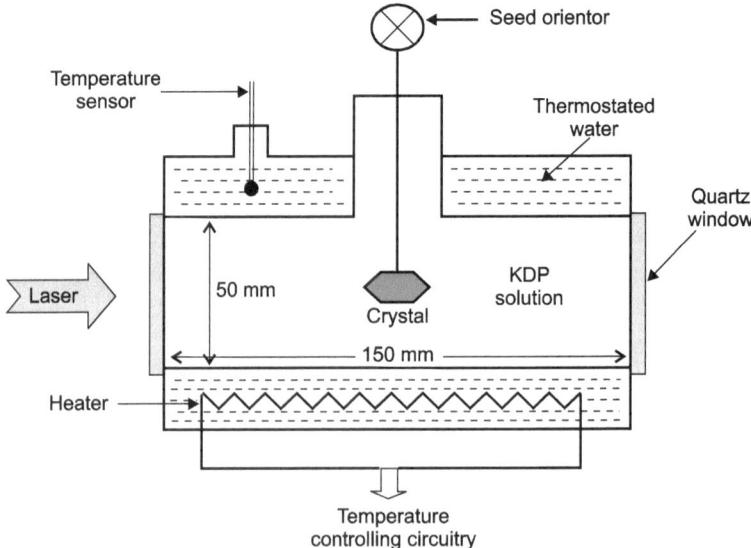

Fig. 4.3 Schematic drawing of the double-walled cylindrical crystal growth cell

chuck is press fitted to the shaft of the stepper motor via a brass bush. A Plexiglas rod of 8-mm diameter, from which the crystal platform is suspended, is tightened into the chuck. The tube section has hollow space machined on either side of it to allow easy placement of the Plexiglas rod. The final assembly was found to rotate without any wobble in the platform.

Growth that starts with spontaneous nucleation is undesirable because it is uncontrollable. Hence, a small crystal of the same material, called a seed, with high physical, mechanical, and optical quality is introduced into the supersaturated solution. Owing to a lower activation barrier, this step would result in preferential crystallization on it rather than at any other location in the crystallizer. The necessity of using a seed crystal can be further understood for those device applications wherein morphology of the crystal required is different from its natural habit of growth. In such situations, a seed of the desired shape, along with appropriate growth parameters, leads to the desired overall crystal shape. Seed crystals are prepared either from spontaneous nucleation (if as-grown morphology is desired in the bigger crystal) or by cutting a smaller portion of the desired shape from a larger crystal. Mechanical handling during cutting of the crystal is crucial because it introduces stresses that act as sources of dislocations in the grown crystal.

A small and good quality seed crystal that is visibly transparent and has the desired morphology is immersed in the solution. The seed is usually 2–3 mm in overall size. At this stage, the solution temperature is kept 0.5 K above its saturation value so as to dissolve a few surface layers of the crystal. Dissolution helps in the elimination of the physical imperfections on the crystal faces and surface contamination introduced during physical handling. After dissolving the crystal for a few minutes, the solution is cooled to its saturation point. Programmed cooling of the solution follows from this point onward and growth over the seed crystal is initiated.

4.5 Experimental Procedure

Results obtained using shadowgraph imaging of the solution around the growing crystal is first presented. Experiments have been performed to include free and forced convection regimes, crystal growth geometries, and cooling rates. As a case study, the growth of a KDP crystal is considered. Results have been analyzed in terms of the development of the convective field around the growing crystal. A linkage between the convection phenomena and the growth rate and crystal quality is examined. The data is quantified in terms of the average as well as instantaneous growth rates along $< 100 >$ and $< 001 >$ crystallographic directions of the KDP crystal. The strength of free convection can be expressed in terms of the Grashof number and forced convection in terms of the Reynolds number. An assessment of the crystal quality is possible in terms of transparency and visible defects such as inclusions, voids, cracks, and striations.

During the experiment, the KDP crystal is grown from a supersaturated aqueous solution prepared using the KDP chemical of 99.5% purity (GR grade, *Merck, India*) dissolved in deionized water that has an electrical resistivity of 18 MΩ-cm (*Millipore,* India). To remove microscopic particulates that cause spurious nucleation, the solution is filtered using a special membrane filter of 0.02-μm pore size and 47-mm overall diameter (*Pall Pharmalab Filtration,* India). The microscopic clusters of solute and microbial impurities present in the solution are eliminated by heating the solution 15 °C above its saturation temperature for 24 h. These two steps considerably help in reducing spurious nucleation.

Crystal growth under free as well as forced convection conditions are discussed in the following sections. The cooling rate of the bulk solution appears as a control parameter. The solution temperature below the initial saturation value is referred as the degree of *supercooling* in the following sections. Additionally, the crystal can be placed inside the crystallizer in various positions. These are (a) crystal hanging from a glass rod, (b) crystal perched on top of a glass rod which in turn is placed on a Plexiglas platform, and (c) crystal directly placed on a platform. Shadowgraph and schlieren imaging of the respective convection patterns are discussed in the following sections.

4.6   Shadowgraph Imaging of Crystal Growth in Free Convection

Buoyancy-driven convection patterns that form around the crystal during growth in various configurations are first discussed in the following sections.

4.6.1   Growth of a Crystal Held on Top of a Glass Rod

A growth experiment is carried out with a lower cooling rate, while the overall geometry is a crystal placed on top of a glass rod. The rod is mounted on a Plexiglas platform. The chosen geometry permits better visualization of the convective plumes over the top face of the growing crystal.

The parameters of growth during the experiment are as follows: A small KDP crystal with bipyramidal morphology having its c-axis horizontal is glued to one end of a thin glass rod (1.5-mm diameter). The glass rod is placed on a Plexiglas platform. Figure 4.4a, b show the schematic drawing and the photograph of the growth geometry. The dimensions of the seed crystal are $2 \times 1 \times 2$ mm^3. The solution has a saturation temperature of 55.2 °C and the average cooling rate adopted during the growth is 0.02 °C/h. The experiment is continued for 195 h of which the initial 45 h are required to reach the saturation temperature and heat the solution to avoid spurious nucleation. Crystal growth continues for 150 h during which the solution is cooled by 3.5 °C.

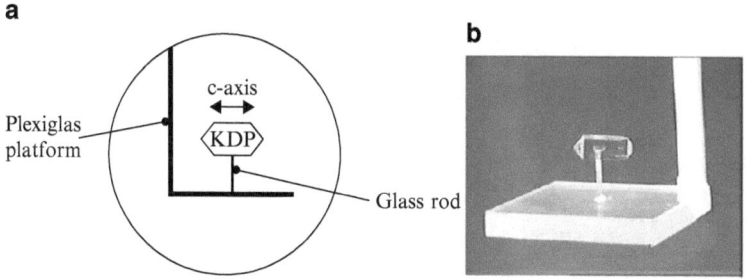

Fig. 4.4 Schematic drawing (**a**) and the photograph of the growth geometry (**b**) for a crystal placed on top of a glass rod

4.6.2 Visualization of Buoyancy-Driven Convection

Figure 4.5a–h show shadowgraph images of the convection patterns recorded during the growth process over a time period of 150 h. The solution is initially saturated and there is no convection seen near to the seed crystal (Fig. 4.5a). The plume gradually builds up as supercooling increases. Initially, the plumes are weak but stable and laminar. This is seen in Fig. 4.5b at a supercooling of 0.5 °C with respect to the initial temperature. After about 60 h of growth, the crystal dimension increases and the plumes become strong. The plumes are seen to rise from the edges of the crystal. The laminar behavior of plumes is maintained at this stage as well, Fig. 4.5b–e. After about 100 h of growth, plumes are seen to emerge from several spots on the top surface of the crystal. This indicates that there are several locations on the top surface where the accumulation of solution depleted of solute act as sources of plumes. At this stage the plume behavior becomes unsteady, Fig. 4.5f, g. Subsequently, the buoyant flow displays chaotic behavior as seen in the irregular plumes of Fig. 4.5h. This is the final state of buoyancy-driven convection observed in the experiment.

The strength of free convection is measured as a function of the degree of supercooling of the solution and characteristic length of the growing crystal by computing the Grashof number

$$Gr = g \left. \frac{\partial \rho}{\partial C} \right|_T \frac{\rho \Delta C H^3}{\mu^2}$$

as given in Table 4.1a. Here, g is acceleration due to gravity, ρ is the solution density, C is the concentration of KDP in the solution, H is the characteristic crystal size, μ is viscosity of the solution evaluated at the average temperature in the growth chamber, and ΔC is the concentration difference driving convection. To a first approximation, it scales with the degree of supercooling of the solution. In the experiments discussed below, Grashof number increases because of a combined effect of an increase in supercooling and an increase in the crystal dimension.

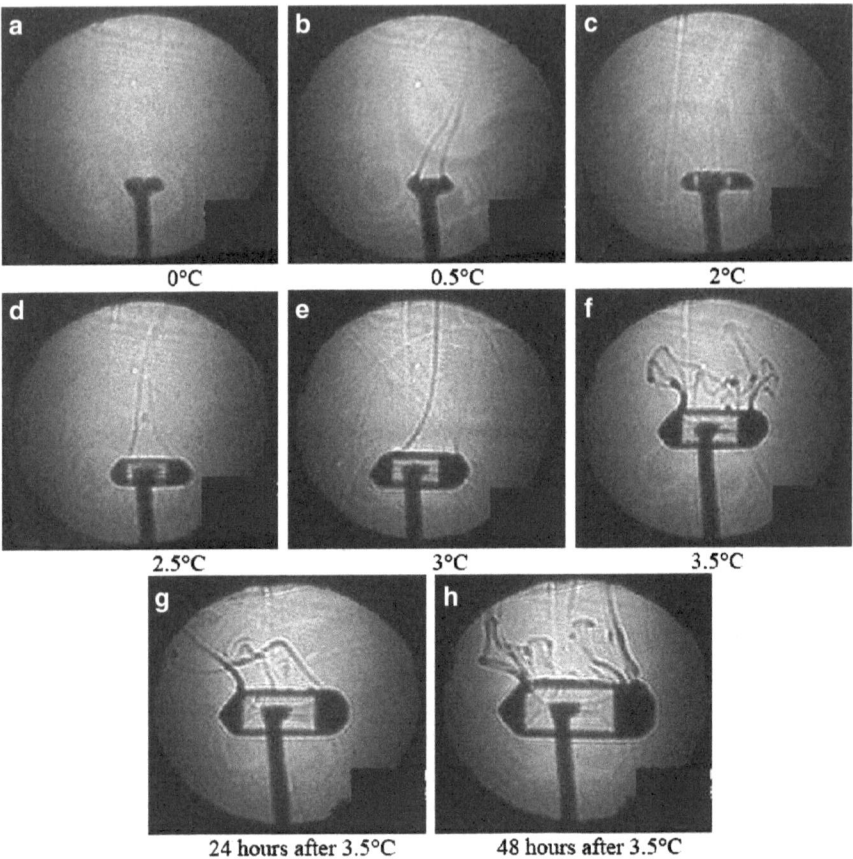

Fig. 4.5 Buoyancy-driven flow transition sequence observed during growth. Initially, as in (**a**), the crystal size is small and the degree of supersaturation is also small, resulting in weak fluid convection. Later, as in (**e**) onward, the crystal size has increased as has the degree of supersaturation and convection is stronger. Under the combined influence of supercooling and crystal size, flow progresses from laminar in the beginning to chaotic at the end, intermediated by irregular flow. Supercooling (in °C) at which the image was recorded is given under each image

Table 4.1 Grashof number as a function of supercooling of the solution and the characteristic length of the KDP crystal. The crystal is placed on top of a glass rod fixed to a platform and grown under free convection conditions

Supercooling (°C)	Characteristic length (mm)	Grashof number (Gr)
0	1.327	0
0.5	1.578	0.79
2	2.136	7.7
2.5	2.702	19.3
3	3.466	48.2
3.5	3.893	78.6

Table 4.2 Growth rate R along $< 100 >$ and $< 001 >$ directions of a KDP crystal as a function of supercooling. The crystal is placed on top of a glass rod fixed to a platform and is grown under free convection conditions. At low supercooling, the growth along $< 001 >$ direction is several times that of $< 100 >$. At higher values, free convection is intensified, and growth is comparable in both directions

Supercooling (°C)	$R < 001 >$ (mm/day)	$R < 100 >$ (mm/day)
0.5	1.257	0.1305
1	2.638	0.254
2	5.262	1.169
2.5	1.553	1.254
3	2.366	1.428
3.5	2.964	0.593

The Schmidt number of the aqueous KDP solution is defined as the ratio of kinematic viscosity and the mass diffusion coefficient. It is around 1,073 and is quite large compared to the Prandtl number of water. For a Grashof number of 100, the corresponding Rayleigh number (= Grashof number × Schmidt number) is 1.073×10^5.

It is observed that Grashof number increases slowly up to a supercooling of 2.5 °C, beyond which it increases sharply. Here, a large degree of supercooling as well as an increase in the crystal dimension plays an important role. At this stage, flow undergoes transition from laminar to chaotic. Once the flow has intensified, the crystal grows rapidly but is prone to generation of defects. It is therefore necessary to have precise control of supercooling to achieve the desired crystal quality while maintaining high growth rate.

During the experiment, the crystal grew from $2 \times 1 \times 2$ to $8 \times 7 \times 17 \, mm^3$ over a period of 150 h. This corresponds to average growth rates of 2.4 and 0.96 mm/day along $< 001 >$ and $< 100 >$ directions of the KDP crystal, respectively. These values are comparable to those reported in the literature for growth under free convection. The instantaneous growth rates along the two directions of KDP crystal as a function of supercooling are given in Table 4.2. It is observed that growth along $< 001 >$ is several times that along $< 100 >$ up to an intermediate value of supercooling (<2°C). However, once the supercooling reaches the stage where free convection is intense, the growth rate along $< 100 >$ decreases. Since the solute available for growth is unchanged, an increase in growth along $< 001 >$ is observed.

Figure 4.6c shows high transparency of the grown crystal. It implies that the cooling rate of 0.02 °C/h is optimum for growing good quality crystal under free convection conditions. Figure 4.6a, b show the crystal located on top of the glass rod that is fixed on a platform and after removal from the platform, respectively. The crystal was clear of any visible defects such as inclusions, bubbles, and streaks except those present in the seed crystal. The percentage transmission in the visible region was measured using a spectrophotometer and was found to be over 80% (with correction for Fresnel losses) revealing good optical quality of the grown crystal.

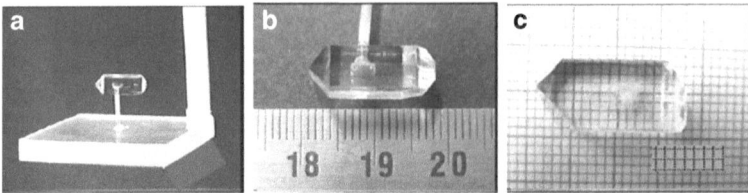

Fig. 4.6 (**a**) Crystal held on top of a rod immediately after removal from the growth chamber, (**b**) an illustration of the size of the crystal, and (**c**) a picture demonstrating the transparency of the grown crystal and its good quality

4.6.3   Growth of a Crystal Placed on a Platform

Apart from the method discussed above, there is a second geometry called platform growth that is suitable for growing large crystals. Shadowgraph imaging has been performed for this geometry to observe the transition sequence of plume behavior. This data can help in establishing a critical range of Grashof numbers beyond which the convection strength becomes detrimental to the crystal quality. For solving the governing equation of refractive index during shadowgraph imaging, the concentration field is usually assumed to be two dimensional. This condition is satisfied only when laminar conditions prevail around the growing crystal. Specifically, free convection patterns need to be recorded from various view angles in order to confirm two-dimensionality of the concentration field. During the experiment, plumes are recorded from several angular directions between 0 and 180 degrees. These images serve additionally as projection data for reconstruction of the 3D concentration field. Under the laminar flow conditions, the plume shape can depend on the crystal geometry. In order to minimize this influence, a seed crystal that is symmetric in shape has been adopted.

The growth parameters during the experiment are as follows. The experiment has been performed under free convection conditions; a symmetrical KDP seed crystal in the shape of a cuboid without pyramidal facets is placed in a small cavity on the platform made of Plexiglas. The dimensions of the seed crystal are $2 \times 2 \times 4\,\text{mm}^3$. The c-axis of the crystal is along the gravity vector, and the crystal is glued to the platform by self-curing silicone sealant that is chemically inert. Figure 4.7a, b show the schematic drawing and the photograph of the growth geometry. The pH of the solution is 4.3, and the impurities in the solution are present in the commercially available KDP chemical of 99.5% purity. The solution has a saturation temperature of $60\,°\text{C}$, and the average cooling rate adopted during the growth is $0.02\,°\text{C/h}$. The experiment is performed for $200\,\text{h}$, cooling the solution by $5\,°\text{C}$.

4.6.3.1   Free Convection Behavior

The shadowgraph images recorded at ten time instants are shown in Fig. 4.8a–j. The first image (Fig. 4.8a) shows the seed crystal in the shape of a cuboid. The

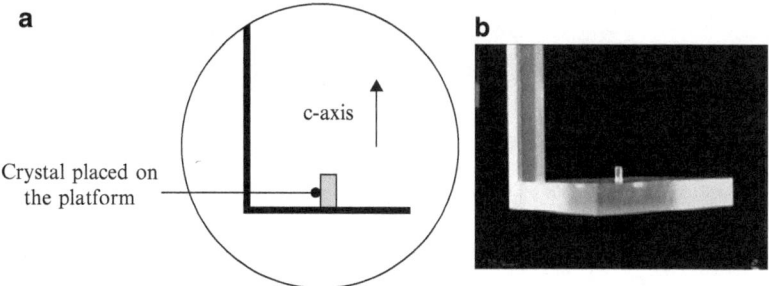

Fig. 4.7 Schematic drawing (**a**) and photograph of the crystal (**b**) at the start of the experiment in platform geometry under free convection conditions. The crystal morphology is a cuboid without any pyramidal facets

second image (Fig. 4.8b) shows the crystal with developed pyramidal faces. In the initial stages of growth, a weak and unstable plume is observed originating from the crystal tip. This plume does not rise far into the bulk and is easily disturbed by stray disturbances in the bulk of the solution. Its frequency of appearance is about 12–15 min. This trend was observed up to a supercooling of 1 °C (Fig. 4.8a–d). With an increase of cooling from 1 °C to 2 °C, the plume frequency increases, and a fresh plume is observed every 5 min. However, it is still weak and unstable in character (Fig. 4.8e, f). When the solution is cooled by 4 °C, the plume strength increases and continuously rises from the tip of the crystal. At this stage, the plume width and its strength increase (Fig. 4.8g). A single plume is replaced by two plumes on either side of the tip of the crystal at a supercooling of 4.5 °C (Fig. 4.8h), which subsequently become three in number (Fig. 4.8i), and finally to five (Fig. 4.8j) for a temperature reduction of 5 °C. This is a clear signature of the increase in the crystal size owing to fluid convection around the crystal.

For a reasonably low degree of supercooling (<2.5°C), experiments show that growth along the < 001 > direction is predominant, resulting in the elongation of the crystal. However, with an increase in supercooling, the growth rates along < 100 > and < 001 > directions get reversed. At this stage, the crystal size increases more along the < 100 > direction than in < 001 >, causing an increase in the crystal cross section. This observation suggests that a controlled cooling rate of ∼0.02°C/h helps in increasing the crystal cross section while preserving crystal quality.

At the end of the experiment, the convective plumes show a pattern of timewise periodicity. The plume is initially regular (Fig. 4.9a–c), changes from laminar to irregular (Fig. 4.9d–f), followed by chaotic (Fig. 4.9g–i), and finally turbulent (Fig. 4.9j). The overall cycle is observed to repeat itself after an interval of a few minutes. This is the transition sequence of free convection for a relatively large crystal.

The strength of free convection at different stages of the experiment (Fig. 4.8a–j) can be quantified in terms of the Grashof number and is given in Table 4.3. For cooling beyond 4.5 °C from saturation, it is seen that Grashof number increases

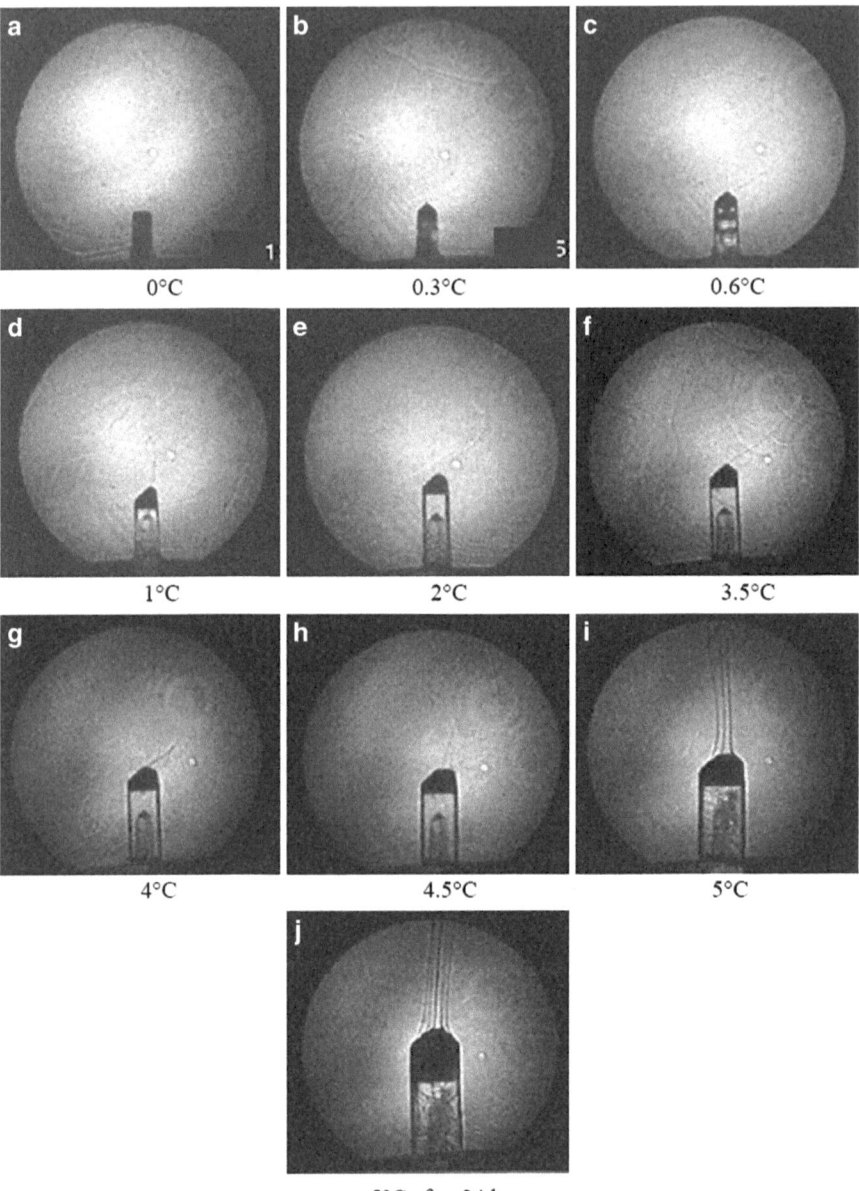

0°C   0.3°C   0.6°C

1°C   2°C   3.5°C

4°C   4.5°C   5°C

5°C after 24 hrs

Fig. 4.8 Plume activity accompanying crystal growth at various stages of supercooling. The degree of supercooling at which the shadowgraph image was recorded is given under each image. Since the growth of the crystal is mostly vertical, buoyant plumes are weak and are seen mainly after stage (**i**)

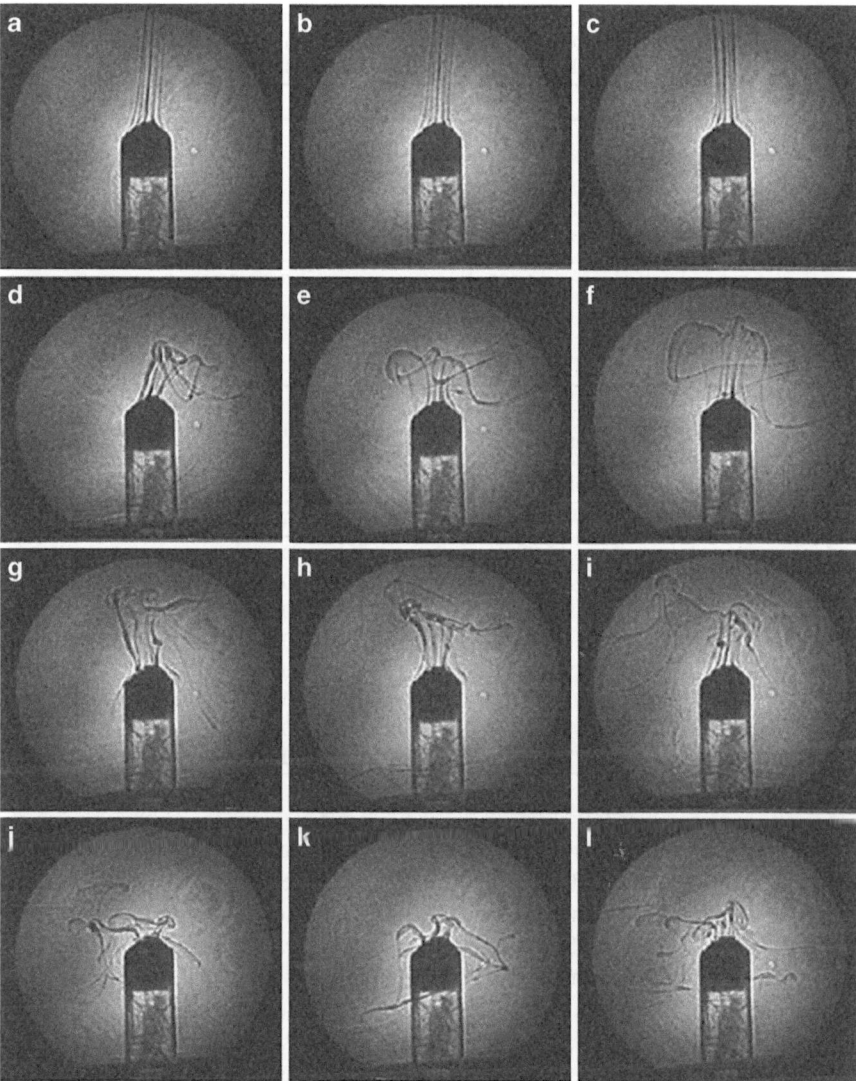

Fig. 4.9 Transition sequence of a plume within a cycle in a grown crystal: Laminar (**a–c**), followed by irregular (**d–f**), chaotic (**g–i**), and finally turbulent (**j–l**). The entire sequence lasts a few minutes, and the cycle is repeated

gradually up to 60. The corresponding convection patterns are steady and laminar. Beyond this value, plume behavior turns irregular and chaotic, and the Grashof number rises rapidly owing an increase in crystal dimension. A Grashof number of 60 can now be interpreted as a critical value at which the buoyant plume becomes unsteady. Below this limit, crystal growth rates are low, while the quality is high.

Table 4.3 Grashof number during successive stages of growth shown in Fig. 4.8. Growth was performed on a platform in free convection conditions. Rayleigh number is 1,073 times the Grashof number

Supercooling (°C)	Characteristic length (mm)	Grashof number
0	1.710	0
0.3	1.781	0.75
0.6	2.098	2.4
1	2.178	4.50
2	2.391	11.8
3.5	2.693	29.0
4	2.916	41.8
4.5	3.214	62.6
5	3.950	127.6

Table 4.4 Growth rate R along $< 001 >$ and $< 100 >$ directions during various stages of growth of a KDP crystal from a platform under free convection conditions

Supercooling (°C)	$R < 001 >$ (mm/day)	$R < 100 >$ (mm/day)
0	0	0
0.3	0.926	0.168
0.6	1.656	0.445
1	1.053	0.050
2	2.01	0.206
3.5	0.062	0.251
4	0.444	0.667
4.5	0.210	0.840

The instantaneous growth rates along $< 100 >$ and $< 001 >$ as a function of supercooling are presented in Table 4.4. The average growth rate of the crystal after the completion of the experiment was found to be 0.5 and 1.35 mm/day along $< 100 >$ and $< 001 >$ directions, respectively. These values are close to those reported in the literature for KDP growth under free convection conditions. The growth rate along the two directions at a given degree of supercooling can be compared with the corresponding Grashof numbers. It is observed that growth rate of $< 001 >$ decreases and that of $< 100 >$ increases at the point where there is a transition to chaotic irregular flow. This can be seen by comparing the data in Tables 4.3 and 4.4. Thus, there is a connection between free convection and the growth rate of the crystal faces. Beyond a Grashof number of around 60, steady–unsteady transition of the plume behavior is observed. Simultaneously, a morphological transition is observed from elongated to an isometric form.

The transparency of the crystal grown under the stable free convection regime with a cooling rate of 0.02 °C/h was seen to be quite good.

4.7 Shadowgraph Imaging of Crystal Growth in Forced Convection

In the above experiments, the crystal growth process is studied under free convection conditions in two different geometries. The primary objective was to map convection around the growing crystal and relate the convective flow patterns to the growth rate and crystal quality. A parameter varied in these experiments was the cooling rate of the solution. Results suggest an optimum cooling rate of 0.02 °C/h for free convection growth. Also, they suggest that the platform geometry is the most favorable, although the growth rate and the quality obtained are comparable to the crystal perched on top of a glass rod. However, one can anticipate difficulties in growing a large crystal perched on a glass rod and a platform geometry is to be preferred.

Extending the growth studies further, a series of experiments under forced flow conditions are performed, keeping the overall geometry, pH, and solutal impurities identical to those in the free convection experiments. Experiments are conducted in the platform geometry. Forced convection is achieved by uniform rotation of the crystal platform in one direction.

4.7.1 Forced Convection by Uniform Rotation

The forced convection experiment is performed by rotating the crystal platform in one direction at a constant rotation rate of 30 RPM. The saturation temperature of the solution is 55.2 °C, and the cooling rate chosen is 0.04 °C/h. A higher cooling rate is required in comparison to free convection to permit high mass fluxes at the surface of the crystal. The crystallizer is a double-walled beaker; a KDP seed crystal with half morphology (single pyramid) and c-axis along the gravity vector is glued to the platform with a chemically inert material. The pH of the solution is 4.3, and the impurities in the solution are present in the commercially available KDP chemical of 99.5% purity. The experiment is performed for 50 h and the total supercooling provided to the solution over this time period is 2 °C. The dimensions of the seed crystal are $1.5 \times 1.5 \times 1.5 \, \text{mm}^3$. Figure 4.10a, b show the schematic drawing and the photograph of the growth apparatus.

A second method of forced convection, called the *accelerated crucible rotation technique*, has been studied by the authors but is not reported here.

With the platform under steady rotation, it is not possible to record the convection patterns. Instead, shadowgraph images are recorded at various stages of growth by stopping rotation for a 1–2 min, Fig. 4.11a–i. Although the crystal grows in the forced convection regime, it is observed during the stationary phase that free convection plumes emerge from the crystal top. These are seen after the swirls in the bulk solution slowly decay. The free convection plumes steadily increase from the initial stages of growth to the final stages of the experiment. The width

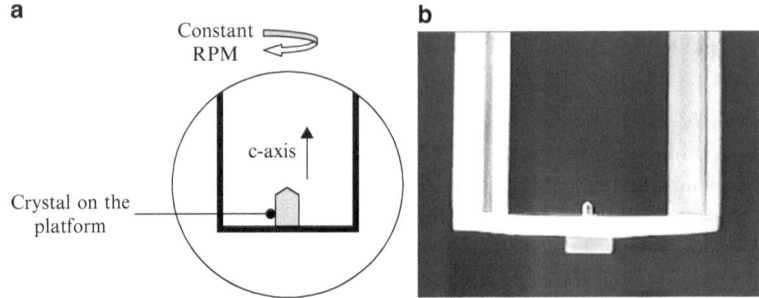

Fig. 4.10 Schematic drawing (**a**) and photograph of the platform geometry (**b**) in forced convection using uniform rotation

of the plume increases with time. Also, the plumes increase in number from one in the beginning to four at the end. The plumes are found to be stable throughout the experiment, becoming irregular intermittently toward the end of the cooling period. After cooling of 1.3 °C is reached, a plume is observed every few seconds and becomes continuous at 1.6 °C. The appearance of stable plumes is a signature of the favorable growth conditions that will yield a good quality crystal. Experiments show that striations appear on the crystal surface when the plume behavior becomes irregular as in Fig. 4.11h–i.

4.7.1.1 Comparison of Growth Rates Along $< 100 >$ and $< 001 >$ Directions

A comparison of growth rates along $< 100 >$ and $< 001 >$ directions of the KDP crystal between experiments in platform geometry under free and forced convection conditions is reported. In case of free convection, it is observed that the growth is predominantly along the $< 001 >$ direction at low cooling (<2.5 °C). When supercooling becomes large, growth along the $< 100 >$ direction increases. To obtain isometric morphology of the KDP crystal, as preferred for device applications, it is imperative under free convection conditions that a high supercooling is provided. Growth under high levels of supercooling is relatively less stable in free convection as compared to growth under forced conditions. This is because of appearance of spurious nucleation in an unstirred solution as opposed to controlled stirring created by platform rotation. Comparing with the morphological evolution of the crystal growing under forced convection as seen in the shadowgraph images (Fig. 4.11a–g) and by obtaining a quantitative estimate of the growth rates along $< 100 >$ and $< 001 >$ directions, it is observed that growth along $< 100 >$ direction takes place even at lower levels of supercooling, and an isometric morphology is obtained. This is an advantage of growth under forced convection. Rotation results in uniformity of the solution concentration, quick removal of the growth debris from the crystal-solution interface, and its replenishment by the solute rich solution. A rapid growth

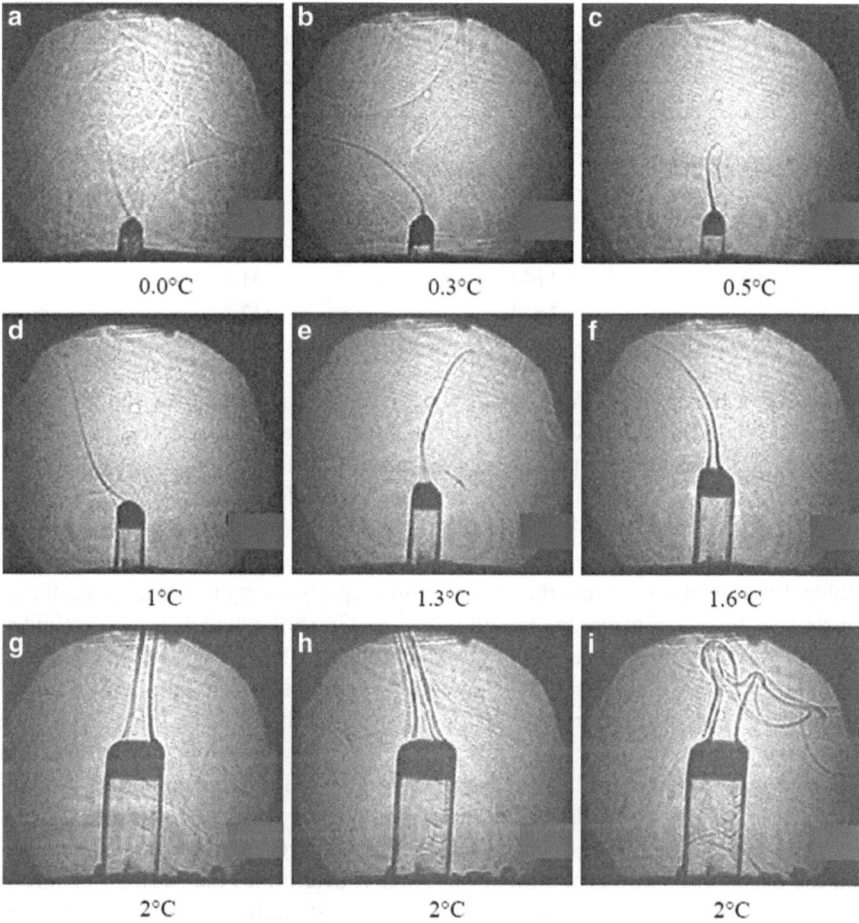

Fig. 4.11 Free convection plumes observed during the stop phase of uniform rotation of the crystal platform. The last three images are at a supercooling value of 2.0 °C where the number of thin plumes increases from 2 to 4 and intermittently become irregular

rate along both the directions is then obtained. Also, no morphological transition occurs because growth rate interchange between $< 001 >$ and $< 100 >$ is not observed at high supercooling, which was seen in free convection experiments. The growth rate along $< 001 >$ remains consistently higher than that along $< 100 >$ for the entire temperature range studied.

4.7.1.2 Strength of Free Convection During the Stop Phase of Rotation

During the stop phase of the rotation cycle, the swirls in the solution decay, and free convection plumes are seen to emerge. These plumes disappear as the fresh cycle

Table 4.5 Grashof number as a function of degree of supercooling and characteristic length for a crystal growing over a platform in uniform rotation. Rayleigh number is 1,073 times the Grashof number

Supercooling (°C)	Characteristic length (mm)	Grashof number (Gr)
0	1.966	0
0.3	2.102	1.13
0.5	2.245	2.290
1	2.687	7.74
1.3	3.067	14.8
1.6	3.683	31.07
2	5.413	115.3

of crystal rotation starts. As a result, the crystal continues to grow under forced convection conditions. However, in order to characterize the free convection activity seen during the brief time periods of the stop phase of the rotation cycle, Grashof number (for solutal buoyancy) has been computed as a function of supercooling and characteristic length at each stage of the experiment. The corresponding Rayleigh number is 1,073 times the Grashof number. The Grashof numbers are given in Table 4.5. It is observed that the crystal grows rapidly owing to higher mass fluxes arising in forced convection. For this reason, Grashof number shows a sudden increase at a relatively lower value of supercooling ($\sim$1.6–2.0°C), when compared to free convection. Conversely, it is possible to apply higher cooling rates in forced convection without deteriorating the crystal quality. It has the additional advantage of a higher growth rate. Thus, forced convection is beneficial for growth but demands a precise control of the cooling rate to avoid unfavorable effects, which may occur during intense flow conditions.

The Reynolds number associated with the growth process needs to be calculated on the basis of the platform dimensions since it determines the strength of forced convection. This value was found to be around 7,685 in the present work, changes being related to that in viscosity with temperature.

The final dimensions of the crystal at the end of the forced convection experiment are $7 \times 7 \times 15 \, \text{mm}^3$. The average growth rate computed along $< 001 >$ and $< 100 >$ directions are 6.48 and 2.64 mm/day. These are significantly higher than those observed in the free convection experiments. The instantaneous growth rates at different stages of growth are presented in Table 4.6. These values also suggest that a many fold increase in growth rate is obtained when compared to free convection in a platform geometry. Also, it may be noted that a higher cooling rate of 0.04 °C/h could be used in comparison to similar experiment under free convection conditions, where the maximum cooling rate was 0.02 °C/h. The quality of the grown crystal in forced convection at the higher cooling rate is excellent, Fig. 4.12a, b.

Table 4.6 Growth rate R along the $< 100 >$ and $< 001 >$ directions as a function of the degree of supercooling for a KDP crystal growing over a platform in uniform rotation

Supercooling (°C)	$R < 001 >$ (mm/day)	$R < 100 >$ (mm/day)
0	0	0
0.3	3.246	0.21
0.5	3.67	1.22
1.0	5.35	2.53
1.3	6.82	1.65
1.6	5.88	2.94
2	7.42	3.80

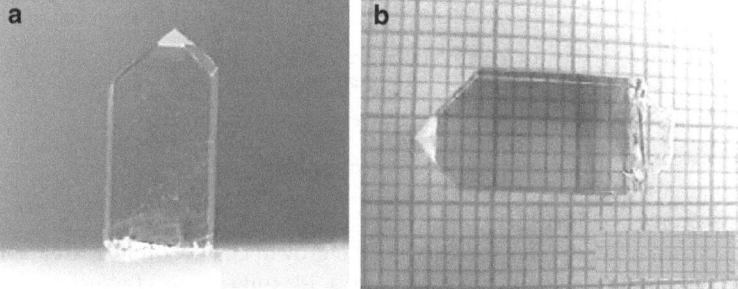

Fig. 4.12 Size and transparency of the crystal grown in platform geometry with constant rotation

4.8   Crystal Growth Chamber for Schlieren Imaging

An independent apparatus was constructed to facilitate schlieren imaging of convection patterns around a KDP crystal growing from an aqueous solution. Apart from recording the intensity data, the images were analyzed for concentration fluxes on various faces of the crystal. The fluxes were integrated to yield the concentration distribution. In the experiments discussed, the cooling rate was large enough to make the solution reach its limit of supersaturation.

The basis for the design of the crystal growth chamber is the volume of KDP solution required to carry out long duration experiments. The chamber is a beaker whose diameter is large enough to minimize the effect of the wall on the strength and orientation of convection currents around the crystal. A large beaker also permits installation of flat optical windows of acceptable size so that a larger area of interest around the crystal can be imaged. The selection of material of the experimental chamber rests on its ability to withstand the operating temperature range of the growth process. It should be smooth enough to prevent nucleation on its surface.

Experiments have been conducted in a growth chamber made of glass (170-mm diameter and 220-mm height) shown schematically in Fig. 4.13. Glass is smooth and chemically nonreactive toward the KDP solution. Moreover, the thermal conductivity of glass is high enough to transmit the gradual changes in temperature

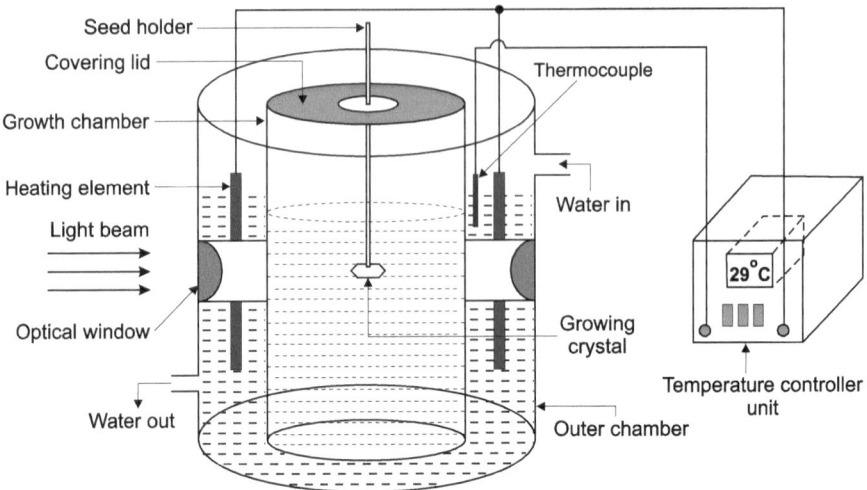

Fig. 4.13 Schematic diagram of the crystal growth chamber

from the thermostated water surrounding the growth chamber to the KDP solution. The crystal growth chamber is placed in a Plexiglas tank. Temperature of the solution inside the growth chamber is maintained at prescribed levels by circulating thermostatically controlled water in the outer tank. The volume of the outer chamber (with dimensions $37 \times 37 \times 26\,cm^3$) is sufficient enough to maintain spatial uniformity of temperature in the solution. The lower surface of the growth chamber rests on a brass base inside the outer tank. Closely spaced holes made on the circumference of the brass base ensure continuous circulation of water beneath the growth chamber as well. Hence, the crystal growth chamber is uniformly in contact with thermostated water. Two heating elements rated at 2 kW are used, while temperature during the cooling process is monitored by a K-type thermocouple fixed on the outer surface of the glass chamber. The temperature of water in the outer tank is controlled to within $\pm 0.1\,°C$ with the help of a programmable temperature controller unit (*Eurotherm*).

For visualization of the concentration field by optical techniques, circular laser-grade optical windows (*Stranda*, BK-7, 60-mm diameter, 5-mm thickness, $\lambda/4$ flatness) are fixed on the walls of the growth chamber at opposite ends. The growth process is initiated by introducing into the growth cell a spontaneously crystallized seed crystal fixed at one end of a thin glass rod. Rotation is imparted with the help of a stepper motor (12 V, 0.6 A, *RS components*, minimum step movement equal to $1.8\,°$/step). A 3-jaw drill chuck is used to connect the seed holder with the stepper motor in order to eliminate the wobbling of the glass rod at high rotational speeds. The initial discussion includes schlieren and shadowgraph images for comparison. Subsequent image analysis is based on schlieren data alone.

4.8.1 Schlieren and Shadowgraph

The growth sequence imaged using schlieren and shadowgraph is shown in Figs. 4.14 and 4.15. The first image in the schlieren and shadowgraph sequences shows dissolution of the seed in the form of a descending plume just after its insertion into the solution. The intensity contrast is related to an abrupt change in the solute concentration around the seed crystal, which creates a jump in the refractive index and deflects light into the region of relatively larger concentration. After initial dissolution, the solution attains thermal equilibrium, and the growth process of the crystal is initiated. Initially, the gradients are small and localized in the vicinity of the growing crystal alone. Growth in the initial stages of the experiment is accompanied by steady, weak convection. Diffusion effects are initially important, and the corresponding intensity contrast in shadowgraph is small. Schlieren images are, however, quite clear. With the passage of time, the size of the crystal increases, convection is intensified, and the concentration gradients grow in strength. This result is brought out in the schlieren images as an increase in the light intensity around the crystal. As defined by the bright region, the resulting flow creates a strong plume directly above the growing crystal. The schlieren image is more vivid (Fig. 4.14), when compared to the shadowgraph (Fig. 4.15). The images also reveal the extent of symmetry of the solutal distribution and the underlying flow field in the stable growth regime of the crystal. For the parameters considered in the experiment, salt concentration in the growth chamber starts to decrease, and the bulk of the solution is stratified.

Schlieren and shadowgraph reveal the following sequence of events: (a) The plumes associated with crystal dissolution are unsteady; (b) the growth process enters a stable growth regime till the crystal size exceeds a critical value; and (c) the flow field approaches a stagnant condition when the solution becomes stratified.

The layering and stable stratification of the solution are seen for times greater than 55 h (Figs. 4.14g, h and 4.15g, h). In the schlieren image, it leads to the region of brightness shifting away from the crystal. Figure 4.14h is an original unprocessed schlieren image. It shows a larger view of convection around the crystal, where stratification leads to a significant refraction and the appearance of a bright patch of light above the crystal. In shadowgraph images, the movement of the stratification front is visible as a bright band that descends vertically downward. For times greater than 50 h, the patch of light is localized around the crystal, while alternating bands of bright and dark regions are formed.

Schlieren (Fig. 4.14c, d) and shadowgraph images (Fig. 4.15e, f) in the stable growth regime show an upward movement of the buoyant plumes around the crystal. These plumes end in the bulk of the solution in the beaker and descend in such a way as to form a closed loop. The concentration gradient in the region away from the crystal is quite small and does not generate sufficient intensity contrast. Hence, the images discussed in the present section pertain to the flux field adjacent to the crystal alone. The buoyant plumes in the stable growth regime are responsible for the deposition of the solute on the crystal surface. These convection plumes are

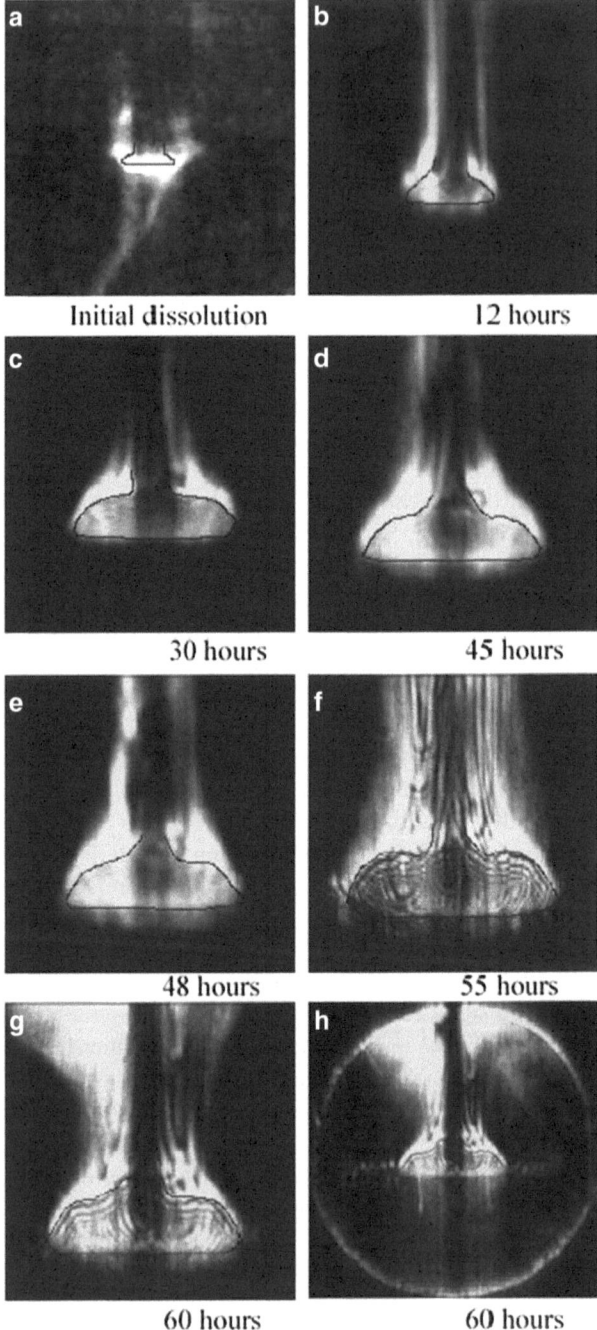

Fig. 4.14 Evolution of schlieren images around the growing crystal from an aqueous solution. In (**h**) the original photograph covering the entire optical window is shown. The crystal position is highlighted

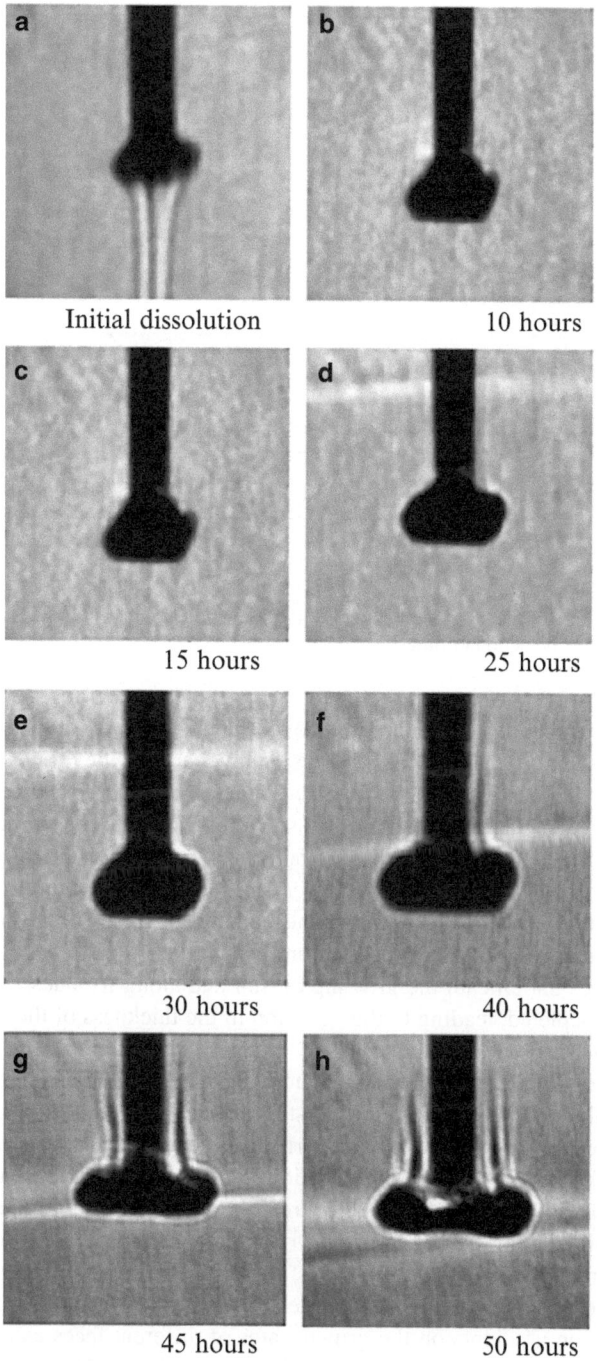

Fig. 4.15 Evolution of shadowgraph images around the growing crystal from an aqueous solution. A bright streak of light indicates the separation of light solution from the heavy. The streak is seen to move downward in (**d**–**h**), till it stabilizes just around the crystal

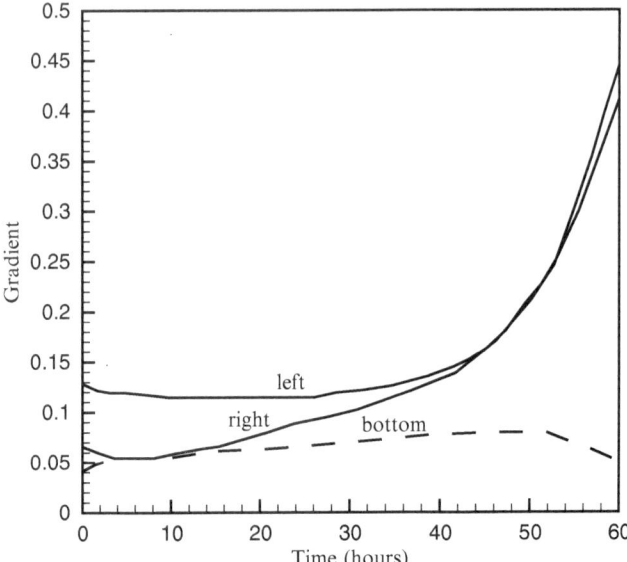

Fig. 4.16 Variation of the dimensionless concentration gradient adjacent to the three faces of the growing crystal as a function of time

nearly steady, resulting in symmetric growth of the crystal. The large gradient of concentration can be visualized as a diffusion layer around the growing faces of the crystal. The thickness of the boundary layer can be identified as the region over which the light intensity is high in the schlieren images. The vertically upward movement of the convective plumes results in the variation of thickness of the concentration boundary layer along the crystal faces. During the growth phase of the crystal, the boundary layer is the thinnest on the lower side of the crystal and the highest on the upper side. This observation can be understood by the fact that as the solution moves underneath the growing crystal and along its sides, the solution is increasingly depleted, leading to the variation in the thickness of the concentration boundary layer.

For a stationary crystal, buoyancy-induced convection is the main driver of solute from the bulk of the solution to the crystal surface. The solution near the crystal being depleted of salt ascends in the vertical direction due to buoyancy and gets replaced by the solution rich in concentration from its vicinity to continue the growth process. With the passage of time, the strength of convection currents depends on the size of the growing crystal and the level of supersaturation of the solution. With an increase in the concentration gradients, convection currents grow in strength that in turn decides the growth rates of the faces of the crystal. The effect of the strength of concentration gradients on the growth rates of different faces can be explained with the help of schlieren images. Figure 4.16 shows the variation of concentration gradients on three different faces (left, right, and bottom) of the growing crystal with respect to experimental run time. As seen in the intensity distribution in schlieren images, the gradients on the side faces grow in strength with time and show similar variation; hence, the two curves (marked *left* and *right*) nearly overlap except in the initial stages of the experiments. The discrepancy for short time can be attributed to the initial unsteadiness in the growth chamber as the rectangular seed crystal transforms into a well-defined KDP structure. Gradients near the lower $< 100 >$ face are comparatively smaller for the entire experimental run. This variation in the gradients is consistent with the schlieren images where bright intensity regions are of higher thickness near the side faces $< 001 >$ as compared to the lower face. Differences in the growth rates of side, top, and bottom faces of the growing crystal can be attributed to the difference in the magnitudes of the concentration gradients. Since the gradients are relatively higher adjacent to the side faces, the horizontal growth rate of the crystal was found to be higher in comparison to the vertical. This observation correlates with the idea of a *dead zone*, a term used to describe an area where growth rates along the $< 100 >$ and $< 010 >$ directions are near zero.

4.9   Control of Convection in a Crystal Growth Process

Mapping of convection patterns as well as the concentration field in the growth chamber is required to establish appropriate conditions for growing large defect-free crystals at meaningful rates. Buoyant convection currents influence the magnitude of the concentration gradients prevailing along the crystal surfaces. In turn, these control the stability of the growth process and the overall crystal quality. The convective field can also be seen as a tool for identifying the role of process parameters. The process parameters studied are cooling rate of the solution, rotational rpm, and the crystal size.

The cooling rate of the solution (the *ramp rate*) determines the amount of excess salt available in the solution for deposition on the crystal surfaces and is the potential difference driving buoyancy-driven convection. Very slow cooling leads to diffusion-dominated growth process and is inefficient; on the other hand, very high values of the ramp rate gives rise to vigorous convection currents resulting in deterioration of crystal quality. In the present set of experiments, a ramp rate of 0.05 °C/h has been employed.

Stirring the solution reduces the natural convection-induced temperature oscillations by homogenizing the bulk solution. Hence, the importance of optimum rates of rotation in crystal growth processes has gained considerable recognition. The two most widely used stirring mechanisms are the rotation of the seed and the rotation of the beaker/crucible. In the present work, the growing crystal is continuously rotated while the beaker is stationary. Crystal rpm of 0 and 15 are studied through experiments. These values have been selected on the basis of their ability to permit growth at acceptable speeds and with meaningful crystal quality.

The crystal size plays an influential role in determining the strength of convection currents in a crystal growth process. With an increase in the length scale, which is related to the change in the size of the crystal, the strength of convection also increases with time till the solution in the growth chamber is depleted of salt. As discussed earlier, this state is characterized by stable stratification of the solution. The growth of the crystal practically stops at this stage. Growth can be resumed when the grown crystal is immersed in a fresh supersaturated solution and the ramp rates are reintroduced.

4.9.1   Ramp Rate of 0.05 °C/h

Figures 4.17 and 4.18 show transient evolution of the convective field when the growing crystal is respectively held stationary in the solution and rotated at a constant speed of 15 rpm. Insertion of the seed into its supersaturated solution can lead to an instantaneous temperature difference between them, followed by an initial dissolution of the crystal. With the passage of time, thermal equilibrium is established, and density differences within the solution are solely due to concentration differences. Deposition of the solute from the solution to the crystal surfaces results in a change of concentration. Locally, the solution goes from being supersaturated to the saturated state. In the absence of rotation, the denser solution displaces the lighter in the vicinity of the crystal, and a circulation pattern is set up around the crystal. With rotation, a radial pressure gradient creates an independent circulation loop that forms an alternative basis of solute movement. Here, the fluid particles around the crystal move in the radial direction, but conservation of mass ensures that vertical velocities be set up once again. In the purely buoyancy-driven mode (0 rpm), the strength and orientation of the convection currents is determined by the available concentration difference in the solution at any instant of time and hence the cooling rate. On the other hand, an externally imparted rotation to the growing crystal leads to homogenization of the solution, reduction in concentration gradients, and hence a reduction in the strength of convection currents. Circulation patterns and their evolution form the basis of solute transport from the bulk of the solution to the growing crystal.

   Figure 4.17 shows the sequence of convection patterns in the purely buoyancy-driven regime. For $t = 20$ h, concentration gradients are primarily localized in the vicinity of the growing crystal. With the passage of time, the size of the crystal increases, and the gradients grow in strength. This result is brought out in the schlieren images as an increase in the light intensity around the crystal. The resulting flow creates a strong plume directly above the growing crystal. Over a longer period of time (20–90 h), the plume structure remains unchanged. It indicates a stable growth regime for the crystal, where buoyant plumes are steady and uniform in nature (Fig. 4.17v–vii). A gradual evolution of the concentration gradients and the associated buoyant plumes ensure a relatively uniform concentration field in the vicinity of the growing crystal, thus leading to symmetric growth of the crystal at the greatest possible rate. As the crystal increases in size, the convection currents grow in strength. Beyond 90 h, they are seen to become quite vigorous

Fig. 4.17 Schlieren images of the evolution of the convective field around crystal growing from its aqueous solution. (Ramp rate = 0.05°C/h, crystal rotation = 0 rpm)

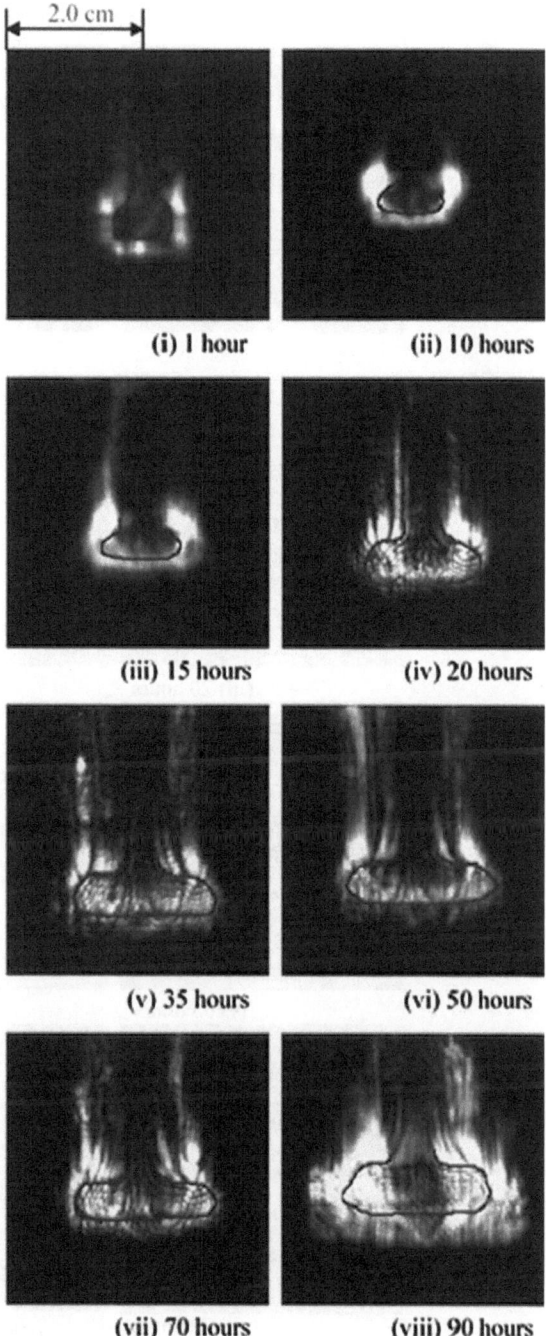

2.0 cm

(i) 1 hour   (ii) 10 hours

(iii) 15 hours   (iv) 20 hours

(v) 35 hours   (vi) 50 hours

(vii) 70 hours   (viii) 90 hours

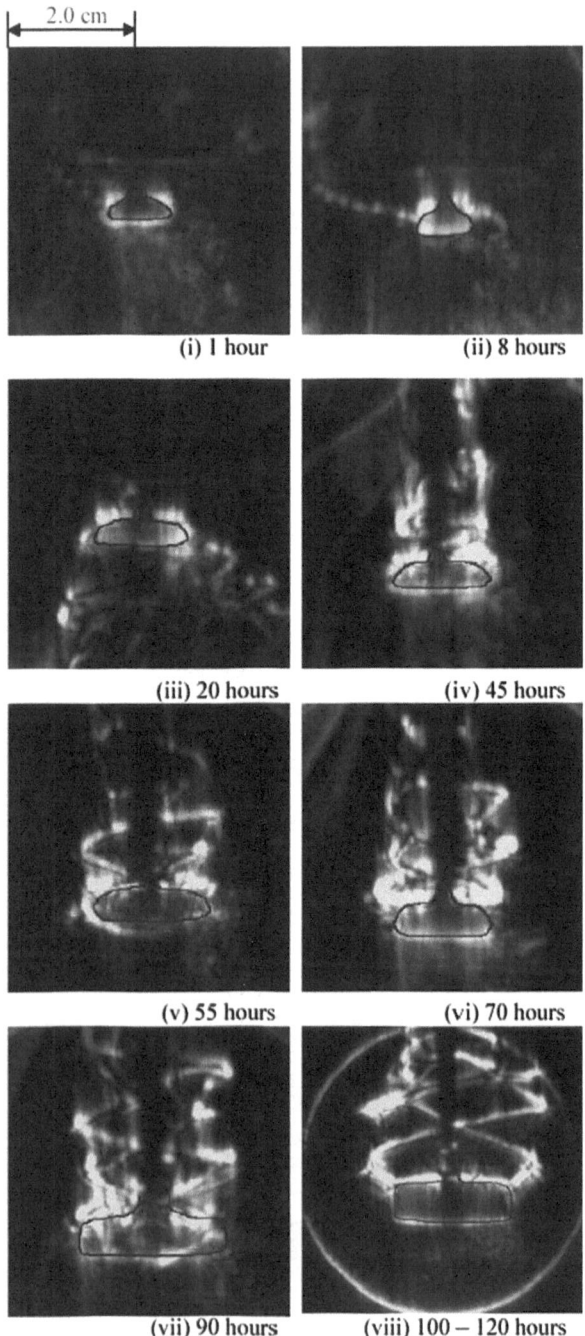

2.0 cm

(i) 1 hour   (ii) 8 hours

(iii) 20 hours   (iv) 45 hours

(v) 55 hours   (vi) 70 hours

(vii) 90 hours   (viii) 100 – 120 hours

Fig. 4.18 Schlieren images of the evolution of the convective field around a crystal growing from its aqueous solution. (Ramp rate = 0.05 °C/h, crystal rotation = 15 rpm)

(Fig. 4.17viii). Correspondingly, time-dependent movement of the plumes was seen in the experiments. This stage is characterized by local changes in the concentration gradients in the vicinity of the growing crystal, followed by a breakdown in the symmetry of the growth process. It is a limit on the time duration for which a single growth experiment can be carried out in the free convection regime. For long times, a stable stratification in density (and hence salt concentration) was obtained in the growth chamber. The rate of increase of the crystal size was negligible at this stage.

Figure 4.18 shows the transient evolution of the convective field around a rotating crystal when the rpm is 15. The first image (Fig. 4.18i) shows the existence of a diffusion layer around the surfaces of the seed crystal, as seen by the nearly uniform distribution of intensity over the growing faces. After a growth period of about 10 h, the gradients near the crystal increase in strength. Unlike the schlieren images in Fig. 4.17 where the buoyant plume moved almost symmetrically along the seed holder over a considerable part of the growth phase, those in Fig. 4.18ii,iii reveal temporary unsteadiness and asymmetric convection currents on either side of the crystal. Unsteadiness can be attributed to (a) temporary unsteadiness in convection due to the development of prismatic faces from the seed crystal and (b) the dominance of centrifugal force over the buoyancy force due to crystal rotation, causing the convection currents to be pushed sideways (Fig. 4.18iii). At subsequent time periods, the combined effect of buoyancy and rotational forces governs the overall orientation and movement of the convection currents.

As the crystal size increases, relatively stronger convection currents rising upward due to buoyancy lead to stronger concentration gradients near its surfaces. The effective movement of the fluid particles is along a helical path, seen in Fig. 4.18iv–vii, but most clearly in Fig. 4.18vii. The width of the helical structure of the rising plumes scales well with the horizontal dimension of the growing crystal. The increasing strength of buoyant convection is evident from the vertical extent upto which the well-defined helical shape is preserved above the crystal. For example, the images for 45 h (Fig. 4.18iv) and 55 h (Fig. 4.18v) show a breakdown of helical structure in the central region between the upper face of the crystal and the free surface of the solution. Helical symmetry is preserved between the crystal and the free surface during the time interval of 90–120 h (Fig. 4.18vii,viii). Hence, at a certain level of supersaturation and crystal size, a delicate balance exists between the rotational and buoyancy forces. The balance provides a geometric pattern to the plume and hence favorable conditions required for the growth of good quality crystals. This particular phase of the growth process with rotation can be termed the stable growth regime in which the crystals of high transparency and symmetry can be grown.

Unlike buoyancy-driven growth, rotation enforces homogenization of concentration around the crystal over the longer duration of experimental run time. Therefore, the possibility of solute stratification as observed in Fig. 4.17viii is delayed. Figure 4.18 shows that the convection regime is purely forced for short time and is governed by crystal rotation ($t = 1$ h). The stirring effect is to be seen by the streaks of light in the image that spread out deep into the solution. Between 1 and 20 h, the spread becomes narrower, as buoyancy forces redirect

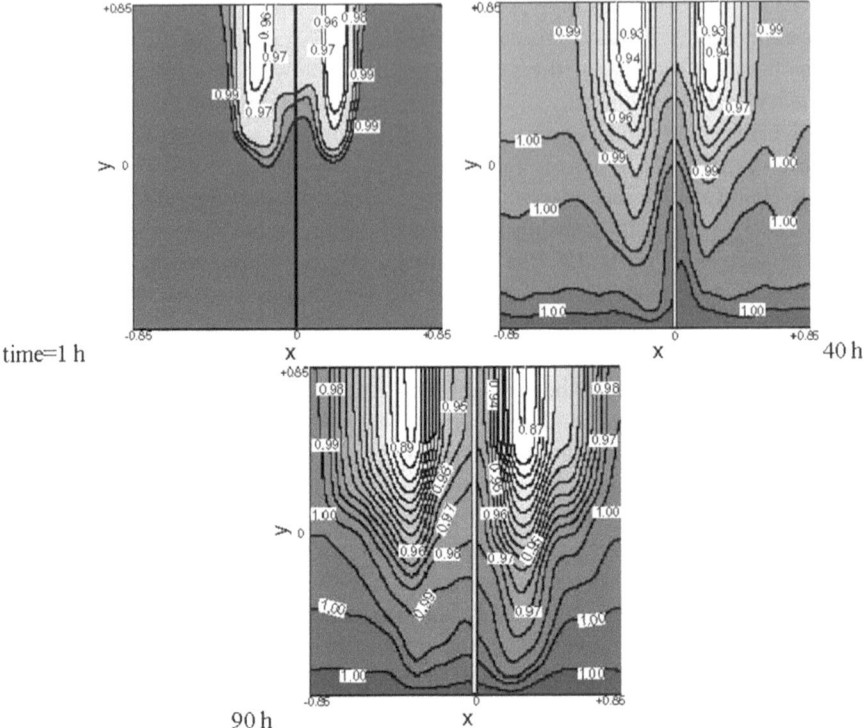

Fig. 4.19 Concentration contours around a growing crystal with the passage of time. The central vertical-filled band in each plot represents the seed holder. (Ramp rate $= 0.05\,^\circ$C/h, crystal rotation $= 0$ rpm)

the plume in the vertical direction. For times greater than 45 h, the plumes show a swirl component but are vertically directed. The relative importance of rotation and gravity is governed by the ratio of buoyancy and centrifugal forces. The force ratio can be shown to be proportional to the crystal size; hence, buoyancy is the guiding force at later times, when the crystal has become large. However, rotation provides a kinematic condition for fluid motion (in the form of a boundary condition), causing the buoyant plumes to become helical and hence structured.

Figures 4.19 and 4.20 show the normalized concentration contour maps around the growing crystal for 0 and 15 rpm, respectively. The value of $C = 0$ represents the saturated state, with $C = 1$ being the supersaturated, both at the temperature of the solution in the growth chamber. The maximum crystal size grown in each experiment has been used to nondimensionalize the x and y coordinates. In the following discussion, the concentration gradient is scaled by the ratio of the maximum concentration difference and maximum crystal size.

The contours in Fig. 4.19 corresponding to zero rotation show a symmetric distribution of the solute in the growth chamber, both near the crystal and in the bulk

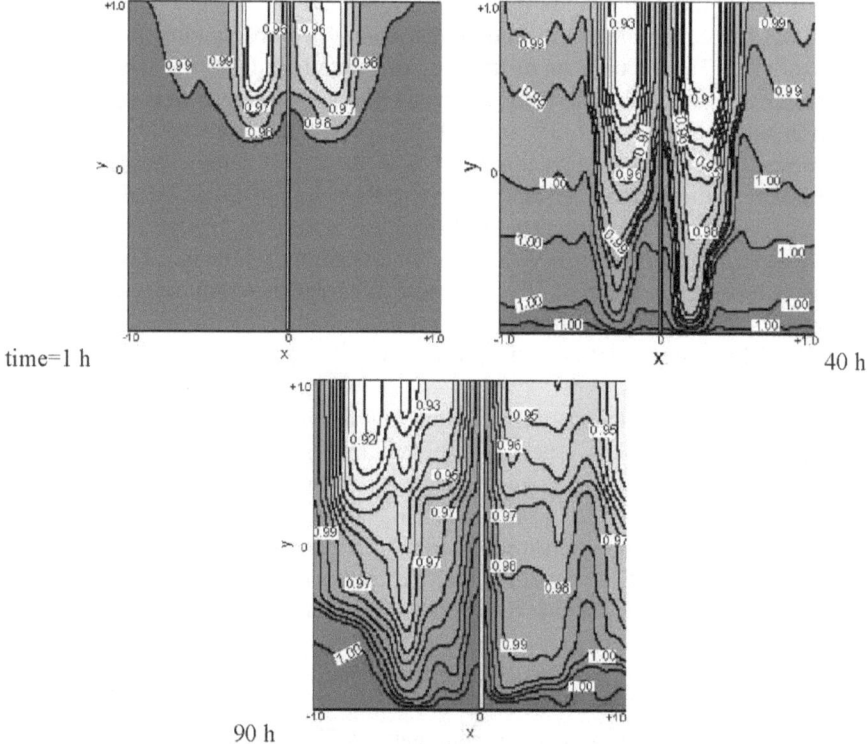

time=1 h   x   40 h

90 h   x

Fig. 4.20 Concentration contours around a growing crystal with the passage of time. The central vertical-filled band in each plot represents the seed holder. (Ramp rate = 0.05 °C/h, crystal rotation = 15 rpm)

of the solution during the initial stages of the growth process, that is, for $t < 10$ h. The contours shown in Fig. 4.20 also reflect a similar trend over the comparable time period. Initially ($t = 1$ h), the contours are localized in the vicinity of the crystal surfaces, with the bulk of the solution being at the supersaturated state ($C = 1$). This phase of the experiment corresponds to the growth process with transport of solute across a diffusion layer to the growing crystal faces. The concentration contours on either side of the seed holder exhibit an asymmetric distribution of solute at $t = 15$ h primarily due to the change in the morphology of the growing crystal from rectangular to prismatic, inducing temporary unsteadiness in the flow field. With the passage of time, the crystal size increases, and the convection currents grow in strength as the growth process enters the stable growth regime. This phase of the experiment is characterized by uniform and structured deposition of solute onto the growing crystal faces. The concentration contours also reveal similar behavior for $t = 40$ h in Fig. 4.19 where symmetric distribution of contours on either side of the seed holder can be seen. The effect of rotation of the growing crystal can be clearly seen in the concentration contours for $t = 40$ h at locations below the crystal

($y = 0$). While those shown in Fig. 4.19 (buoyancy-driven) depict a larger spread in the upward direction along the seed holder, contours corresponding to 15 rpm are relatively flat. Thus, a uniform distribution of solute is indicated in the solution in the presence of rotation. The vertically rising plume is narrower in the presence of rotation but broadens with time. At $t = 90$ h, Fig. 4.19 shows a slight breakdown in symmetry of the concentration contours below the crystal, with respect to the seed holder, though the plume maintains symmetry. Rotation, on the other hand, enforces a better symmetry to the concentration distribution around the crystal. This is clearly visible at 90 h in Fig. 4.20. At 90 h, the plume is strongly influenced by buoyancy and is as broad as in the case of no rotation. The swirl flow pattern superimposed on buoyant flow by crystal rotation imparts anti-symmetry to the isoconcentration contours. This pattern continues to represent a new form of symmetry for the growth of good quality crystals.

The growth rate of the crystal is related to concentration gradients, rather than concentration alone. This aspect is explored in Fig. 4.21. The variation of the dimensionless concentration gradient averaged over each of the three faces of the growing crystal (left $< 001 >$, right $< 001 >$, and base $< 100 >$) with respect to experimental run time are shown in Fig. 4.21a for 0 and 15 rpm. As expected, left–right symmetry of the crystal is realized in growth with and without rotation. The effect of rotation is to lower the overall concentration gradient when compared to that generated by buoyancy alone. The gradients on the lower face are small in comparison to the sides. On the lower face, density stratification is stable, and buoyant motion is inhibited. The effect of rotation is then to increase the gradients here by inducing fluid movement. Evolutionary profiles on the top face of the crystal are not shown, since the image is affected by the presence of the seed holder. In the purely buoyancy-driven experiment, the gradients on the side faces grow in strength with time, whereas the gradients along the lower face are small. The increase in the gradients on the side faces is consistent with the corresponding high intensity regions in the schlieren image sequence of Fig. 4.17. The problem of high concentration gradients during the later stages of experiments ($t \geq 60$ h) and also a significant difference in the relative distribution of these gradients over sides and lower faces of the growing crystal is seen to be overcome by rotation. The effect of rotation in equalizing the strength of the gradients over the three faces of the crystal is indicated by the closeness of the gradient profiles in Fig. 4.21a. Figure 4.21b shows the horizontal growth of the crystal with respect to the experimental run time. The growth rate with rotation is slightly lower when compared to buoyancy alone; it is however practically linear. The growth rate with crystal rotation is comparatively lower because of two factors: (a) the lowering of concentration gradients in the vicinity of the growing crystal due to homogenization of the solution induced by crystal rotation and (b) the rotation of the crystal introduces a radial (outward) velocity component that inhibits the transport of solute to its growing surfaces.

Figure 4.21c shows photographs of the crystals grown with and without rotation. The size of the grown crystal (after 90 h) is larger in buoyancy-driven convection, but the crystal quality is superior in terms of transparency when growth is accompanied by rotation.

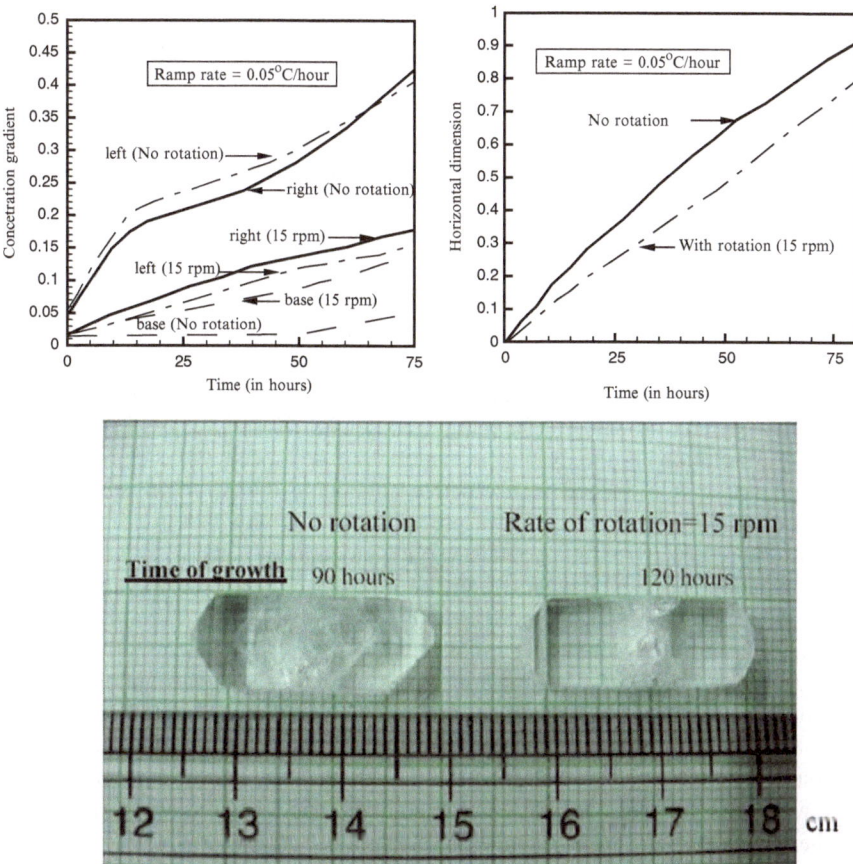

Fig. 4.21 Variation of nondimensional concentration gradients near the three faces of the growing crystal (**a**), horizontal growth of the crystal relative to its maximum size as a function of time (**b**). (Ramp rate = 0.05 °C/h). (**c**) Photographs of grown crystals with and without crystal rotation

Measurements in situ, related to the surface quality of the crystals growing from solution, can also be carried out using optical techniques [8, 10]. Joint imaging of convection and surface quality has been discussed by the authors elsewhere.

4.10   Growth of Protein Crystals

While large KDP crystals are required in photonics applications, protein crystals of suitable composition and uniform size are required in directed drug delivery systems and several biomedical contexts [1,2,4]. These crystals are much smaller than those of KDP but larger than what would be obtained in an uncontrolled crystallization process. Large protein crystals are also required to study the protein structure itself. Such crystals can be grown from solution by controlled evaporation near-ambient temperature. The crystal size and quality would depend on the resulting mass transfer phenomena [6, 7, 11, 12, 14–16, 19]. Protein crystal growth from its solution in a suitable medium is the subject of the following discussion.

Proteins can be prompted to form crystals under appropriate conditions [9]. One approach is that the purified protein undergoes slow precipitation from an aqueous solution. As a result, individual protein molecules align themselves in a repeating series of unit cells by adopting a consistent orientation. The crystalline lattice formed is held together by noncovalent interactions. The goal of crystallization is to produce a well-ordered crystal, free of contaminants. It should be large enough to provide a clear diffraction pattern during the passage of X-rays. The diffraction pattern can be analyzed to discern the three-dimensional structure of the protein.

Protein crystal growth in a hanging drop configuration is discussed here. The drop is a solution of the lysozyme material in deionized (DI), demineralized, distilled water. It is placed in a crucible containing the *reservoir solution* that is an aqueous medium containing a precipitant along with a buffer. The drop as well as the reservoir solution contain varying amounts of NaCl dissolved in them. The combination is then sealed off from the atmosphere. As the drop evaporates, the solution in the drop is supersaturated with lysozyme that, in turn, precipitates out as crystals.

According to Raoult's law, the vapor pressure of a solution containing a non-volatile solute such as NaCl is equal to the vapor pressure of the pure solvent (water) at its temperature multiplied by its mole fraction. Thus, in a dilute solution, the vapor pressure is higher. During protein crystal growth, the salt concentration of the drop is lower when compared to the reservoir solution, and hence the vapor pressure is higher. In this respect, the difference of concentration of NaCl in the drop compared to the reservoir generates the potential difference for solvent evaporation from the drop. Water vapor leaving the drop condenses as fresh H_2O over the reservoir solution. In turn, salt concentration gradients are created in the reservoir, leading to a mass diffusion phenomenon. Within the drop, the solution becomes supersaturated with respect to lysozyme that precipitates in crystalline form. The crystal growth process continues till the drop has fully evaporated, leaving a collection of lysozyme crystals of various sizes adhering to the surface. Parameters that govern the speed of the growth process and crystal quality are salt concentration, volume of the reservoir solution, initial drop volume, and the pH of the reservoir solution.

Evaporation of water from the drop involves absorption of thermal energy and hence lowering of the liquid temperature. However, the entire process is quite slow, and one can assume that isothermal conditions prevail within the sealed crucible.

4.10.1   Protein Crystallization Chamber

Attempting to crystallize a protein without a proven protocol can be a tedious process. Factors that require consideration are protein purity, pH, concentration of the buffer and precipitants such as NaCl, and the overall temperature. The protein should be at least 97% pure. Variation in pH values can result in distinct crystal structures. A buffer such as sodium acetate is necessary for maintaining the pH of the solution. Precipitants include ammonium sulphate and sodium chloride and cause the protein to precipitate out of solution. Polyethylene glycol is often used as a cryoprotectant. Sodium azide (NaN_3) is an antimicrobial preservative that is used to protect samples and crystallization reagents from microbial contamination at weight concentrations of 0.02% to 0.1%. An experimental apparatus for the growth of lysozyme crystals combined with optical access for laser imaging is described below.

Lysozyme crystals are grown in the laboratory scale in circular small wells that share a cover plate made of polystyrene, an optically clear plastic material. It is not convenient to work with this apparatus when optical imaging with a parallel beam of light is to be performed. Hence, a larger rectangular cavity has been fabricated for the present study.

The basis for the design of the crystal growth test cell is clear visualization of the diffusion fields in the gap between the drop and the reservoir as well as within the reservoir solution. It is also of interest to see the reduction in drop radius with time. The selection of material for the experimental chamber rests on its ability to be nonreactive with the chemical solutions used, transparency, surface smoothness, and also a low thermal conductivity. In the present work, 2-mm-thick Plexiglas sheet has been used to fabricate the test cell. The overall dimensions of the apparatus are $127 \times 31 \times 31$ mm^3. The schematic diagram of the test cell is shown in Fig. 4.22. The faces of the apparatus that permit the passage of the laser beam are made of optical grade glass (BK-7). Important precautions for the growth experiment include careful sealing of the apparatus and thermal insulation from the ambient so that room temperature fluctuations do not influence the measurements.

The rectangular cavity is filled up to a predetermined height with the reservoir solution of known composition. A drop of lysozyme solution prepared in DI water mixed with the reservoir solution is placed on the stopper in an inverted position over the reservoir to achieve the *hanging drop* configuration [20, 22, 24, 27]. Optical measurements in the form of interferograms, schlieren, and shadowgraph have been carried out, though the discussion below is in only terms of monochrome and color schlieren images. In most experiments, the intensity contrast in the gas phase between the drop and the reservoir was found to be small. In addition, the drop curvature precluded imaging of any transport within the drop itself. Hence, the results below pertain mainly to mass transfer of NaCl within the reservoir, as the fresh condensate of H_2O mixes with the rest of the solution. Since fresh water is lighter than the reservoir solution, the stratification is gravitationally stable and convection currents are not generated. A time sequence of images is recorded to

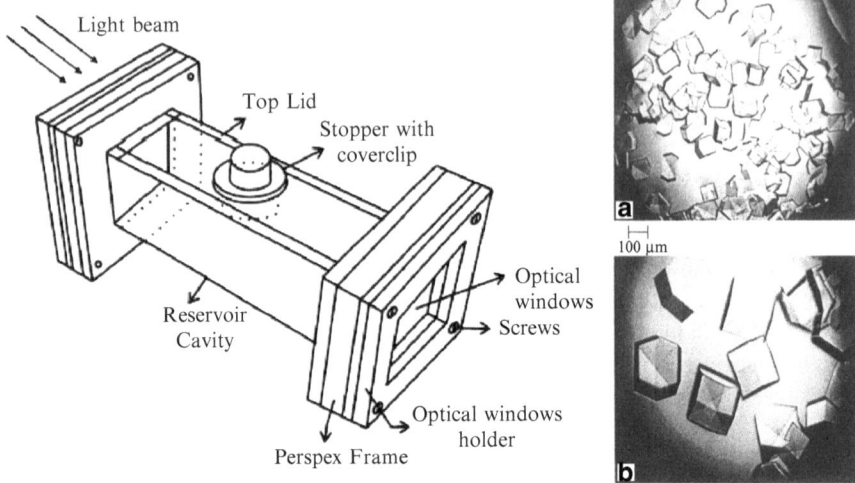

Fig. 4.22 Schematic diagram of a rectangular cavity for studying mass transfer during protein crystallization. A photograph comparing controlled and uncontrolled (*top*) crystallization of lysozyme in a hanging drop arrangement is shown below the experimental apparatus. The photograph has been recorded through a stereomicroscope

understand the mass transfer process that occur during the crystal growth process. In the results presented below, the drop solution is a mixture of the protein and the reservoir solutions in a ratio given as $p : r$. While the reservoir volume is changed, its composition is held fixed at 30% (w/v) NaCl solution, 70% (v/v) ethylene glycol solution, and sodium acetate buffer (0.1 M, pH $= 4.8$).

4.10.2 Color Schlieren

The evolution patterns of moisture diffusion imaged using color schlieren technique are respectively shown in Fig. 4.23. The thin black patch at the interface between air and the reservoir solution is due to meniscus formation that blocks the light beam in the viewing direction. The first in the sequence in color schlieren is the base image obtained soon after placing the drops in the growth chamber. Water evaporation from the drop is then initiated. Water vapor condenses over the reservoir creating a region free of NaCl. Figure 4.23b shows a greater amount of fresh water in the central portion of the interface after 0.5 h, directly below the drops. Condensation continues while fresh water diffuses into the reservoir. Mass diffusion of moisture in the reservoir generates its own spread of colors in the interface region (Fig. 4.23c–e).

The diffusion process causes fresh water to migrate vertically downward. Hence, a gradual change of color is seen in the downward direction. The color differential in color schlieren is related to an abrupt change in salt concentration of the reservoir

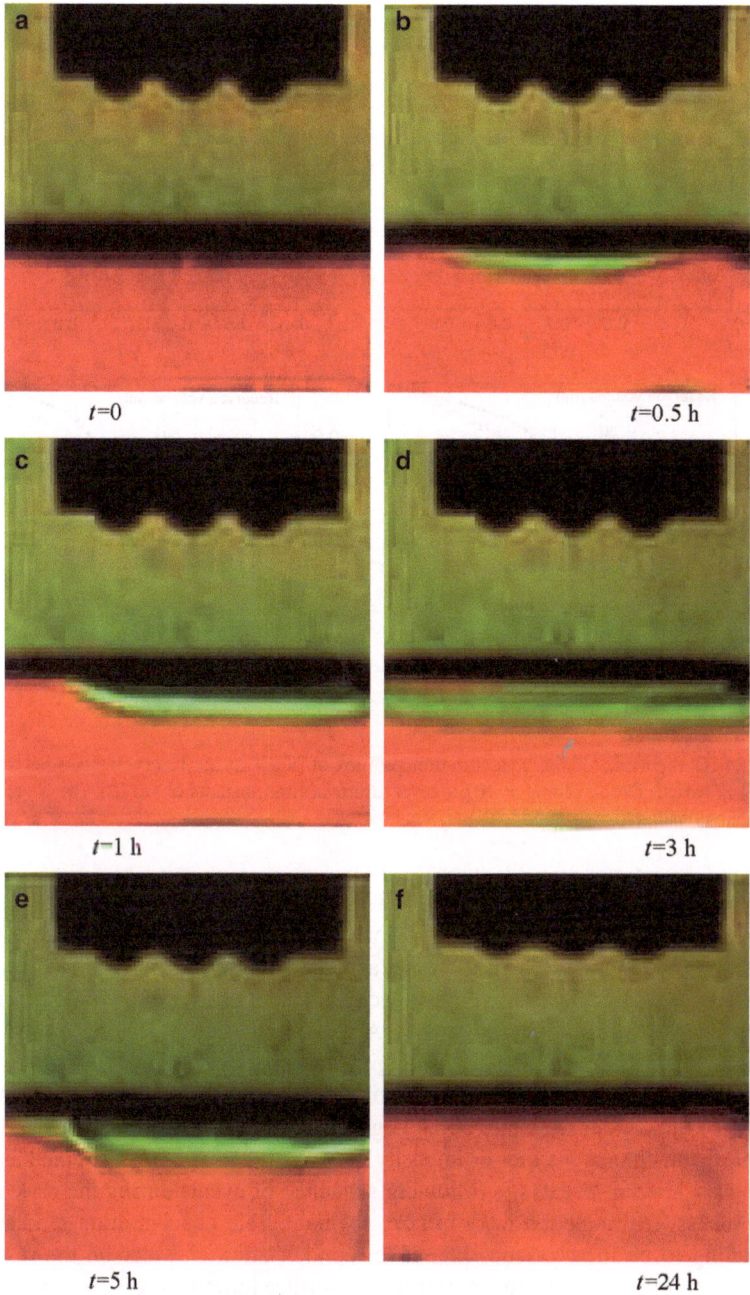

Fig. 4.23 Color schlieren images of transient evolution of diffusion process in reservoir during lysozyme protein crystal growth. Seven protein drops with drop concentration $p : r = 7 : 3$ and 50 ml reservoir solution are used

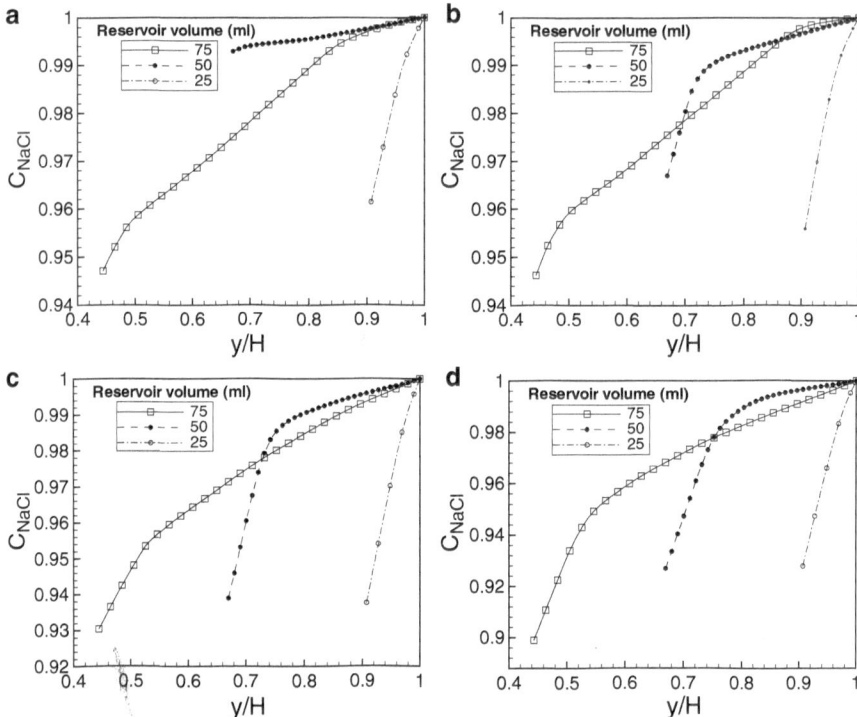

Fig. 4.24 Comparison of the concentration profiles of NaCl in the reservoir solution when its volume is varied. Drop volume = 10 μl; seven drops at time instants of (**a**) 0.5, (**b**) 1, (**c**) 2, and (**d**) 4 h are considered

solution with respect to the deposited water. It creates a jump in the refractive index and deflects the light beam into the region of relatively large concentration. In the portion of the cavity between the drop and the reservoir, color changes in color schlieren are small. This intervening space is, hence, not very significant from the view point of controlling crystal growth. After 24 h, the color schlieren image looks similar to the base image. The drops hanging underneath the top surface have nearly disappeared at this time indicating the completion of the evaporation process. The volume of the condensed water being negligible compared to the reservoir, there is no significant change in color or intensity between the final image and the base.

Color schlieren reveals the following sequence of events during the crystallization process: (a) reduction in drop size, (b) negligible concentration gradients in air leading to an inert gaseous phase, (c) condensation of water on the reservoir surface, (d) diffusion of fresh water from the surface into the reservoir solution, and finally (e) the appearance of a well-mixed reservoir solution of nearly uniform NaCl concentration. Concentration profiles of NaCl that reveal steps (d) and (e) are shown in Fig. 4.24. Here, the distance measured from the base of the reservoir is denoted as y, and H is the height of the rectangular cavity. The data has been generated, first in terms of the concentration gradient and followed by integration from the base of the reservoir where the concentration field is undisturbed. The condensation of fresh water over the reservoir lowers the salt concentration. The depth of penetration of fresh water changes with time. In the experiments considered, up to 40% of the reservoir depth is seen to be affected. The second observation is that the salt concentration near the interface diminishes with time over a period of nearly 2 h. This phase comprises diffusion of clear water into the solution. For later times, the salt fluxes are directed toward the interface, and the local concentration increases.

4.11 Summary

Schlieren and shadowgraph images of solutal convection patterns around a KDP crystal growing from an aqueous solution are discussed. Convection plumes, mass fluxes, and crystal growth rates are seen to be controllable by means of the ramp rate and crystal rotation. The mass fluxes diminish with rotation but lead to an improvement in crystal quality. Signatures of protein crystal growth rates are seen in gradients of NaCl in the reservoir. These gradients respond strongly to the salt concentration difference between the drop and the solution as well as their respective volumes. The growth process is diffusion dominated.

References

1. H. Adachi, K. Takano, M. Morikawa, S. Kanaya, M. Yoshimura, Y. Mori, and T. Sasaki, Application of a two-liquid system to sitting-drop vapour-diffusion protein crystallization, Acta Crystallographica, Section D Biological Crystallography, Vol. D59, pp. 194–196, 2003.
2. N. Asherie, Protein crystallization and phase diagrams, Methods, Vol. 34, pp. 266–272, 2004.
3. I. Braslavsky and S.G. Lipson, Interferometric tomography measurement of the temperature field in the vicinity of a dendritic crystal growing from a supercooled melt, Trans. of optical methods and data processing in heat and fluid flow, IMeChE, London, pp. 423–432, 1998.
4. N.E. Chayen, and E. Saridakis, Is lysozyme really the ideal model protein? J. Crystal Growth, Vol. 232(1–4), pp. 262–264, 2001.
5. W. C. Chen, D. D. Liu, W. Y. Ma, A. Y. Xie and J. Fang, The determination of solute distribution during growth and dissolution of $NaClO_3$ crystals: the growth of large crystals, J. Cryst. Growth, Vol. 236, pp. 413–419, 2002.
6. A.A. Chernov, J.M. Garcia-Ruiz, B.R. Thomas, Visualization of the impurity depletion zone surrounding apoferritin crystals growing in gel with holoferritin demer impurity, J. Crystal Growth, Vol. 232, pp. 184–187, 2001.
7. V.K. Chhaniwal, A. Anand, C.S. Narayanamurthy, Measurement of diffusion coefficient of transparent liquid solutions using Michelson interferometer, Optics and Lasers in Engineering, Vol. 42(1), pp. 9–20, 2004.
8. K. Onuma, T. Kameyama and K. Tsukamoto, In-situ study of surface phenomena by real time phase shift interferometry, J. Crystal Growth, Vol. 137, pp. 610–622, 1994.
9. A.S. Parmar, P.E. Gottschall and M. Muschol, Pre-assembled clusters distort crystal nucleation kinetics in supersaturated lysozyme solutions, Biophysical chemistry, Vol. 129, pp. 224–234, 2007.

10. E. Piano, G.A. Dall'Aglio, S. Crivello, R. Chittofrati and F. Puppo, A nondestructive interferometric technique for analysis of crystal growth and fluid dynamics, Ann. Chem. Sci. Mat. Vol. 26, pp. 23–28, 2001.

11. F. Pullara, A. Emanuele, M.B. Palma-Vittorelli and M.U. Palma, Lysozyme crystallization rates controlled by anomalous fluctuations, J. Crystal Growth, Vol. 274(3–4), pp. 536–544, 2005.

12. J. Qi and N.I. Wakayama, Solute convection during the whole process of protein crystal growth, J. Crystal Growth, Vol. 219, pp. 465–476, 2000.

13. L. N. Rashkovich and G.T. Moldazhanova, Growth kinetics and bipyramid-face morphology of KDP crystals, in *Growth of Crystals*, pp. 69–78, E.I. Givargizov and A.M. Melnikova editors, Consultants Bureau, New York, 1996.

14. R. Savino and R. Monti, Buoyancy and surface-tension-driven convection in drop protein crystallizer, J. Crystal Growth, Vol. 165, pp. 308–318, 1996.

15. G. Sazaki, T. Matsui, K. Tsukamoto, N. Usami, T. Ujihara, K. Fujiwara and K. Nakajima, In situ observation of elementary growth steps on the surface of protein crystals by laser confocal microscopy, J. Crystal Growth, Vol. 262, pp. 536–542, 2004.

16. A.M. Schwartz and K.A. Berglund, In situ monitoring and control of lysozyme concentration during crystallization in a hanging drop, J. Crystal Growth, Vol. 210(4), pp. 753–776, 2000.

17. I. Sunagawa, K. Tsukamoto, K. Maiwa and K. Onuma, Growth and perfection of crystals from aqueous solution: Case studies on barium nitrate and K-alum, Prog. Crystal Growth and Charact., Vol. 30(2–3), 153–190, 1995.

18. I. Sunagawa, *Crystals: Growth, Morphology, and Perfection*, Cambridge University Press, UK (2005).

19. M. Takehara, K. Ino, Y. Takakusagi, H. Oshikane, O. Nurekib, T. Ebina, F. Mizukami and K. Sakaguchi, Use of layer silicate for protein crystallization: Effects of Micromica and chlorite powders in hanging drops, Analytical Biochemistry, Vol. 373, pp. 322–329, 2008.

20. R.E. Tamagawa, E.A. Miranda and K.A. Berglund, Simultaneous monitoring of protein and $(NH_4)_2SO_4$ concentrations in a protein hanging-drop crystallization using Raman spectroscopy, Crystal Growth & Design, Vol. 2(6), pp. 511 -514, 2002.

21. B. Vartak, A. Yeckel and J. J. Derby, Time-dependent, three-dimensional flow and mass transport during solution growth of potassium titanyl phosphate, J. Crystal Growth, Vol. 281, pp. 391–406, 2005.

22. P. G. Vekilov and F. Rosenberger, Protein crystal growth under forced solution flow: experimental setup and general response of lysozyme, J. Cryst. Growth, Vol. 186, pp. 251–261, 1998.

23. S. Verma and P.J. Shlichta, Imaging techniques for mapping solution parameters, growth rate, and surface features during the growth of crystals from solution, Progress in crystal growth and characterization of materials, Vol. 54, pp. 1–120, 2008.

24. J.M. Wiencek, New Strategies for Protein Crystal Growth, Annual Review of Biomedical Engineering, Vol. 1, pp. 505–534, 1999.

25. W. R. Wilcox, Transport phenomena in crystal growth from solution, Prog. Crystal Growth Charact., Vol. 26, pp. 153–194, 1993.

26. A. Yeckel, Y. Zhou, M. Dennis and J. J. Derby, Three dimensional computations of solution hydrodynamics during the growth of potassium dihydrogen phosphate II. Spin down, J. Crystal Growth, Vol. 191, pp. 206–224, 1998.

27. D.C. Yin, Y. Inatomi, H.M. Luo, H.S. Li, H.M. Lu, Y.J. Ye, and N.I. Wakayama, Interferometry measurement of protein concentration evolution during crystallization and with improved reliability and versatility, Measurement Science and Technology, Vol. 19, pp. 1–8, 2008.

Chapter 5
Imaging Jet Flow Patterns

5.1 Introduction

Several authors [1–4, 10, 16] have reported jet flow behavior in distinct flow configurations. Imaging jet flow in various contexts is the subject of this chapter. Configurations such as (a) buoyant helium fountain, (b) multiple jets and (c) jet impingement are included. These examples illustrate the utility of optical techniques in explaining interesting flow patterns. Quantities such as penetration distance, spreading rates, and instabilities are discussed. Schlieren and shadowgraph images are included.

5.2 Buoyant Jets

Schlieren images of a negatively buoyant jet passing through nozzles of circular and square cross sections are presented here for a wide range of Froude and Reynolds numbers. The flow configuration comprises a helium jet emanating from a tube in the vertically downward direction into a normal ambient. For the range of dimensionless parameters studied, steady and unsteady flow patterns were both realized. In these experiments, pure helium was used as the working fluid. Schlieren images were recorded with various orientations of the knife-edge. A horizontal knife-edge that cuts off the upper portion allows a better view of the fountain cap. A vertical knife-edge with the side cut off shows the depth of penetration of the jet. The initial setting of the knife-edge was such as to block off 100% of the laser beam, leading to a dark background. The flow field has thus been recorded with superior contrast. Densimetric Froude number Fr and Reynolds number Re, calculated using the tube radius as the length scale and the space-averaged tube exit velocity as the velocity scale, are

$$Re = \frac{UR}{\nu} \quad Fr = \frac{U}{\sqrt{g'R}}.\tag{5.1}$$

P.K. Panigrahi and K. Muralidhar, *Imaging Heat and Mass Transfer Processes*, SpringerBriefs in Applied Sciences and Technology 4, DOI 10.1007/978-1-4614-4791-7_5, © Pradipta Kumar Panigrahi and Krishnamurthy Muralidhar 2013

Table 5.1 Summary of experimental conditions employed for imaging a buoyant jet

Nozzle shape	Nozzle diameter	Flow rate (in lpm)	Froude number	Reynolds number
Circular	4 mm	0.75	2.85	17.12
		1	3.80	22.83
		2	7.60	45.67
		3	11.40	68.50
		4	15.19	91.33
Circular	5 mm	1	2.17	18.27
		2	4.35	36.53
		3	6.52	54.80
		4	8.70	73.07
Circular	6 mm	1	1.38	15.22
		2	2.76	30.44
		3	4.13	45.67
		4	5.51	60.89
		5	6.89	76.11
		6	8.27	91.33
		7	9.65	106.56
Square	6 mm	1.5	1.62	17.93
		2	2.16	23.91
		3	3.25	35.86
		4	4.33	47.82
		5	5.41	59.78

Here, g' denotes the effective acceleration due to gravity that accounts for the change in the gas mixture density ρ_m with respect to air (of density ρ_a) and is given as

$$g' = g \times \frac{\rho_m}{\rho_a}.$$

Froude number is set within the range of 1–18, while the corresponding Reynolds numbers fall in the range of 15–125.

The nondimensional parameters in the experiment were set by adjusting the flow rate and the nozzle diameter. Increase in Froude number is accompanied with that in Reynolds number. The two parameters relate the importance of inertia against gravity (for Froude number) and viscous forces (for Reynolds number). Quantities of interest in this chapter are penetration length and oscillation frequency, both of which scale strongly with Froude number. Hence, the discussion below is in terms of Froude number alone. To examine the controllability of the jet, nozzles of various shapes have been used. For a circular nozzle, the internal diameter varies over 4–6 mm. For a square nozzle, the hydraulic diameter was 6 mm. Table 5.1 provides experimental conditions employed for imaging. The data obtained has been critically examined for repeatability. Images recorded in a buoyant fountain are discussed in the following sequence.

1. Visualization images
2. Penetration length

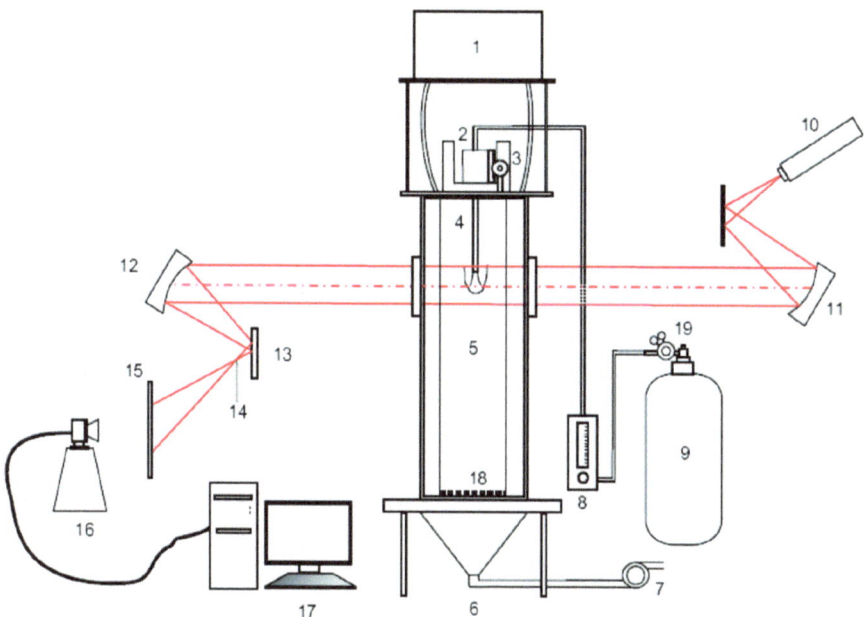

Fig. 5.1 Schematic drawing of the experimental setup. *1*. Gas collection chamber, *2*. nozzle housing, *3*. traverse, *4*. nozzle, *5*. test cell, *6*. exit region, *7*. purging blower, *8*. rotameter, *9*. gas cylinder, *10*. He-Ne laser, *11*. collimating mirror, *12*. decollimating mirror, *13*. flat mirror, *14*. knife-edge, *15*. viewing screen, *16*. camera, *17*. image acquisition system, *18*. wire mesh, and *19*. gas flow regulator

3. Power spectra
4. Strouhal number

Data recorded in the present studies have been validated where available against published results for a buoyant helium jet.

5.2.1 Experimental Apparatus

The schematic drawing of the experimental setup with laser schlieren arrangement is shown in Fig. 5.1. The laser schlieren system includes a He-Ne laser, concave mirrors, flat mirrors, knife-edge, and the image acquisition system. The time-averaged flow rates are obtained from the fluid-specific rotameters (*Scientific Devices*).

The flow facility consists of a gas cylinder (9), gas flow regulator (19), rotameter (8), gas collection chamber (1), nozzle housing (2), nozzle (4), and test cell (5) as shown in Fig. 5.1. Starting from the helium cylinder, flow is controlled by a pressure regulator (19), and flow rate measured using a helium flow rotameter (8).

Depending on the experimental requirement, two rotameters with their individual flow measurement limits have been used. The respective flow ranges are 0.5–5 lpm and 2–20 lpm of helium with least counts of 0.25 and 1.0 lpm, respectively. A metered amount of helium enters a plenum with a square cross section attached to the nozzle housing (2). The gas passes through antiturbulence screens and a long nozzle (5) that ensures fully developed flow at the exit of the nozzle.

Two nozzle geometries with round and square cross sections have been used to understand buoyant jet behavior. Since a fully developed velocity profile is required at the nozzle exit, the L/R ratio needs to be large. For creating a circular jet, seamless copper tubes of diameter 4, 5, and 6 mm and 180-mm length have been used. The square tube of 6-mm hydraulic diameter and 200-mm length has been fabricated in the laboratory. The test cell gets filled with helium–air mixture during the experiment. It is periodically purged so as to maintain uniform ambient conditions within the test cell.

The test cell is made with Plexiglas sheets (12-mm thickness). It has an octagonal cross section, 1 m long, and the distance between parallel faces equals 216.7 mm. Two square optical windows (45×45 mm) are mounted on the sides of the test cell for the passage of the laser beam. Continuous flow of helium leads to an unfavorable accumulation of gas at the top of the test cell and limits the duration of the experiment. This effect has not been considered in previous studies (e.g., [8] and [33]). In the present work, the test chamber was periodically purged, and in addition, the test cell had passages for continuously withdrawing helium. The apparatus shown in Fig. 5.1 is quite similar to the one described in [5] in terms of orientation and [21] in terms of inflow conditions. Papanicolaou et al. [24] have pointed out that a large variation is possible in the measurement of penetration length owing to confinement effects. These authors have recommended the plate separation to be four times the jet diameter.

5.2.2   Flow Visualization

Laser schlieren images of a helium fountain, created by a helium jet directed downward, is presented for a range of dimensionless parameters. For schlieren imaging, both positions of the knife-edge—vertical and horizontal—are considered. For a vertical position, horizontal density gradients are highlighted, while the horizontal knife-edge emphasizes vertical gradients. Visualization images show that the flow is inertia dominated for small times but buoyancy forces appear later and a steady state is reached at low flow rates. At higher flow rates, the longtime behavior may involve oscillations. The discussion below is in terms of the starting behavior, steady state, and unsteady characteristics.

He ↓

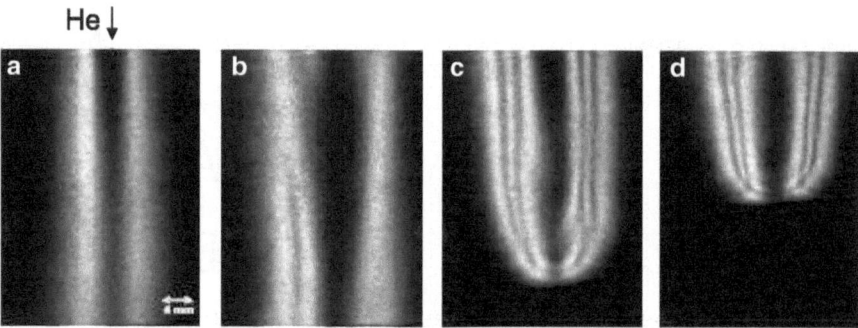

Fig. 5.2 Schlieren visualization of the starting behavior of a buoyant helium jet with a horizontal knife-edge setting ($Fr = 6.52$, $Re = 54.80$, $R = 2.5$ mm); time instants considered are (**a**) 0.4, (**b**) 1, (**c**) 2, and (**d**) 7 s. Steady state was reached in about 7 s

5.2.2.1   Starting Behavior of the Jet

Starting behavior of Boussinesq and non-Boussinesq forced plumes are reported in [11]. A time sequence of schlieren and shadowgraph images of a buoyant helium jet is shown in Fig. 5.2, with $Fr = 6.52$ and $Re = 54.80$. The main flow direction is vertically downward, with buoyancy forces on the jet acting in the upward direction. The jet penetration in the ambient is the highest for small time but decreases with time and is the smallest at steady state. The small time behavior of the jet is analogous to positively buoyant jet discussed in [12] and [18]. The starting behavior of the jet is seen at time $t = 0.4$ s in Fig. 5.2a. The overshoot in the jet position is compensated by buoyant forces that decelerate the flow. The penetration length then starts to decrease. The gas particles that have reached near-zero vertical velocity move sideward, and a cap is formed. The radial velocity field now permits a plume formation with the gas moving upward in the annular region around the main jet. The entire phenomenon is affected by the transport of helium in air since the mixture density is a function of the gas concentration. In addition, buoyant forces depend on the density difference, being larger in regions of large helium concentration and near zero in the ambient. The transport process by which the helium mixes with air has a convection component related to entrainment and a diffusion component. It is to be expected that the former is significant when the jet-ambient velocity difference is large, for example, at early times as well as in the momentum-dominated portions of the jet at later times. Diffusion is important in the stagnant portions of the gas in contact with the ambient, for example, around the cap. Compared to thermal diffusivity of gas, mass diffusivity of helium in air is particularly large. Hence, the buoyant fountain of a helium jet has distinct characteristics when compared to a hot air fountain. The penetration length, transients, as well as longtime fluctuations depend on the extent of mixing of helium with air and are discussed in the following sections.

In previous theoretical and experimental studies on salt-water fountains [31], the initial penetration length was 140–150% of the steady value. Philippe et al. [25]

Fig. 5.3 Sketch depicting
flow regions A, B, and C in
the proposed model of a
helium fountain

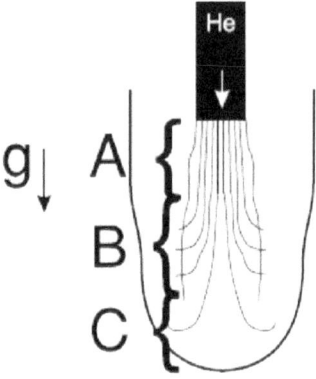

attributed the difference between steady state penetration depth and maximum penetration depth to the large jet-ambient density difference. However, there is no experimental evidence to support this claim. In case of a helium fountain, the difference can be as large as 1,200%. This peculiar behavior of helium fountain can be attributed to lack of surface tension at helium-air interface and a large mass diffusivity of helium in air. The second factor characterizes helium as a non-Boussinesq fluid with its unique spreading properties.

Along the lines of [5], a new model, shown in Fig. 5.3, is proposed for a buoyant helium jet mixing with air in a vertically downward orientation. Three distinct regions are identified as follows:

1. *Region A* represents the momentum-dominated region in which continuous growth of the jet is expected due to entrainment from the environment. Here, buoyancy-induced annular upflow surrounds the inner flow. High shear is expected at the periphery of the inner core, but the jet is mainly transported downstream. Figure 5.2a shows a distinctly visible flow in the downward direction for short times. The annular flow is related to the returning jet fluid, moving vertically upward due to a lower density relative to the ambient fluid. Apart from velocity differences, the mass fraction of helium in the inner core is greater than that of the annular fluid and a strong diffusive mass transport mechanism is set up. Hence, due to the radial concentration gradient, diffusive mass transfer and convective mass transfer are equally important in the outer region.

2. *Region B* is buoyancy affected with some influence of inertia. The vertical gas velocity has diminished in the inner core owing to buoyancy. In this region, entrainment is weaker, and gas from the inner core spreads radially outward by diffusion. Figure 5.2b is an appropriate time instant when the jet is in region B. Here, the effect of diffusive mass transfer in the radial direction is the most dominant transport mechanism.

3. *Region C* is the cap of the fountain in which fluid particles have practically come to rest, inertia effects being a minimum, and buoyancy, the most dominant factor affecting fluid motion. The spherical shape of the cap is related to the centerline

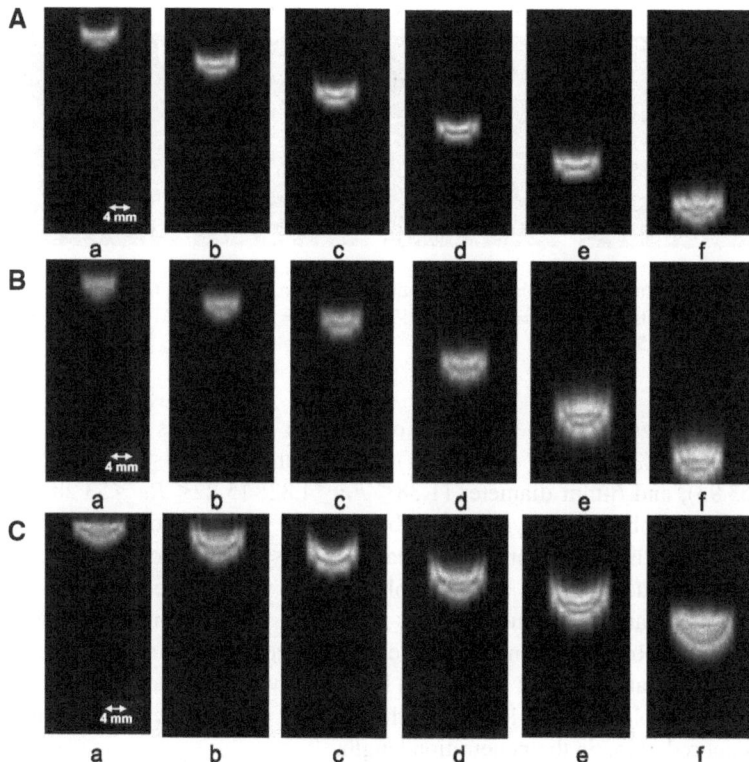

Fig. 5.4 Steady state schlieren images of the helium fountain created by round nozzles of various internal radii for various flow conditions: (**A**) $R = 2.0$ mm: (**a**) $Fr = 3.80$, $Re = 22.83$; (**b**) $Fr = 5.70$, $Re = 34.25$; (**c**) $Fr = 7.60$, $Re = 45.66$; (**d**) $Fr = 9.50$, $Re = 57.08$; (**e**) $Fr = 11.40$, $Re = 68.50$; (**f**) $Fr = 13.30$, $Re = 79.91$; (**B**) $R = 2.5$ mm: (**a**) $Fr = 2.17$, $Re = 18.27$; (**b**) $Fr = 3.26$, $Re = 27.40$; (**c**) $Fr = 4.35$, $Re = 36.53$; (**d**) $Fr = 5.44$, $Re = 45.67$; (**e**) $Fr = 6.52$, $Re = 54.80$; (**f**) $Fr = 7.61$, $Re = 63.93$; (**C**) $R = 3.0$ mm: (**a**) $Fr = 1.38$, $Re = 15.22$; (**b**) $Fr = 2.07$, $Re = 22.83$; (**c**) $Fr = 2.76$, $Re = 30.45$; (**d**) $Fr = 3.45$, $Re = 38.06$; (**e**) $Fr = 4.13$, $Re = 45.67$; (**f**) $Fr = 4.82$, $Re = 53.28$

velocity being greater in Poisseuille flow that affords greater transport along the axis in the vertically downward direction. Since helium accumulates within the cap, concentration differences with the surrounding fluid make diffusive mass transfer the most dominant transport mechanism in this region.

McDougall [17] has proposed a similar model identifying regions A and C for turbulent jets of a Boussinesq fluid. This model is incomplete for a helium jet in air since it has a large mass diffusivity, clearly giving rise to region B for the range of parameters studied.

5.2.2.2 Steady State Characteristics

Figure 5.4 shows schlieren images obtained for a round nozzle after steady state has been reached. Schlieren images have been recorded with the knife

He ↓

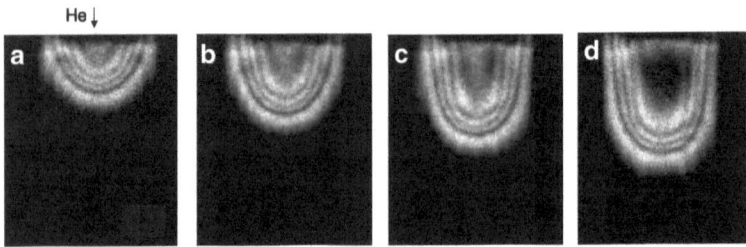

Fig. 5.5 Steady state schlieren images of a helium fountain created by a 6-mm square nozzle for various flow conditions: (**a**) $Fr = 1.62$, $Re = 17.93$ (**b**) $Fr = 2.16$, $Re = 23.91$ (**c**) $Fr = 2.71$, $Re = 29.89$ (**d**) $Fr = 3.25$, $Re = 35.87$

edge kept horizontal. The three sets of data are for nozzles of 4-mm diameter ($3.80 \leq Fr \leq 13.30$; $22.83 \leq Re \leq 79.91$), 5-mm diameter ($2.17 \leq Fr \leq 7.61$; $18.27 \leq Re \leq 63.93$), and 6-mm diameter ($1.38 \leq Fr \leq 4.82$; $15.22 \leq Re \leq 53.28$). In each set, Froude number and Reynolds number increase from left to right. Experiments showed that the helium fountain was steady for $Re \leq 36$. The increase in penetration depth with Froude number and Reynolds number is consistently seen in all the images. The change in Froude number is stronger for a given nozzle diameter in comparison to Reynolds number. Hence, the increase in penetration length can be associated primarily with an increase in Froude number. With an increase in nozzle diameter, Reynolds number increases while Froude number decreases, along with a consistent reduction in the penetration length.

With reference to the model shown in Fig. 5.3, one can conclude that region A is predominant at small Froude numbers, while the three regions A–C are realized at large Froude numbers.

Figure 5.5 shows schlieren images for a helium jet streaming from a 6-mm square nozzle. The Froude and Reynolds number ranges are $1.62 \leq Fr \leq 3.25$ and $17.93 \leq Re \leq 35.87$. Schlieren images for a square nozzle are similar to the round nozzle of equal diameter. In view of the large diameter, Froude numbers are small, and region A is dominant in the flow field (Fig. 5.6). Correspondingly, the schlieren image is mostly the cap of the jet where vertical fluid velocities are small and the gas diffusion effects with the ambient are strong. At the highest Froude number studied, the schlieren image shows annular flow surrounding the inner jet, namely, region B of the model proposed in Fig. 5.3.

5.2.2.3 Unsteady Characteristics

Instability and breakdown of variable density jets are reported in the literature [13, 22, 30]. With increasing Froude number, buoyancy forces become strong, the jet forms a cap adjacent to the nozzle, and the gas is directed vertically upward. For a given Froude number, fluctuations appear in the flow field for an increase in Reynolds number. A correlation proposed in the literature [33], namely,

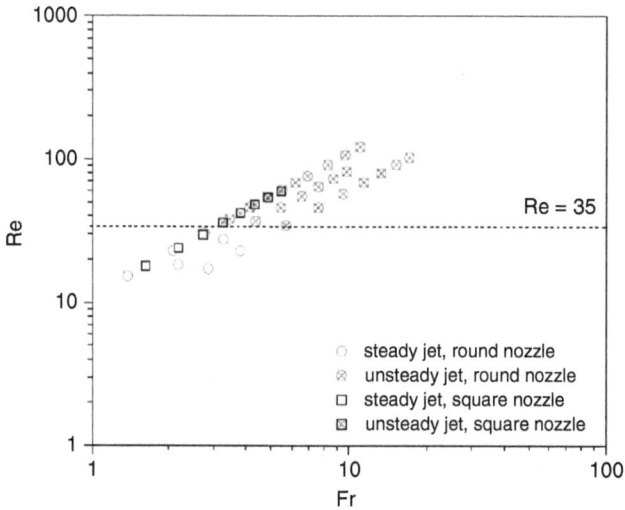

Fig. 5.6 Regime map of fountain behavior with *Re* and *Fr* as coordinates. Initiation of time-dependent fluctuations is observed at *Re* = 35

$$Fr \times Re^{0.67} = \text{constant}$$

demarcates initiation of fluctuations in a laminar jet and its progress toward various modal patterns. This dependence on Froude number is not seen in the helium experiments. A reason for this behavior could be ascribed to the special transport properties of helium. Specifically, it has a kinematic viscosity that is ten time larger than air, making Reynolds number the appropriate dimensionless parameter for characterizing the onset of instability.

Post instability, the mode shapes of jet oscillation (Figs. 5.7–5.10) depend on Froude number as well as Reynolds number. These shapes are classified as follows:

1. *Bobbing*: This behavior is seen as an oscillation in the fountain height. It is clearly perceived in schlieren as movement of the jet cap. Oscillations in the jet fountain height under turbulent conditions were studied by [8] and [24]. The bobbing behavior in helium jets is seen under laminar conditions and is comparable to that reported in [33].
2. *Flapping*: It refers to the sideways fluctuation of the fountain, while the penetration length is practically constant. Flapping behavior can be seen in both positions of the knife-edge—horizontal as well as vertical (Fig. 5.9).
3. *Flapping + Bobbing*: With increase in Reynolds and Froude numbers, fluctuations are intensified, and jet oscillations are seen in vertical as well as horizontal directions (Fig. 5.7). The two motions may alternate in time, or be mixed chaotically, in distinct portions of the parameter space.

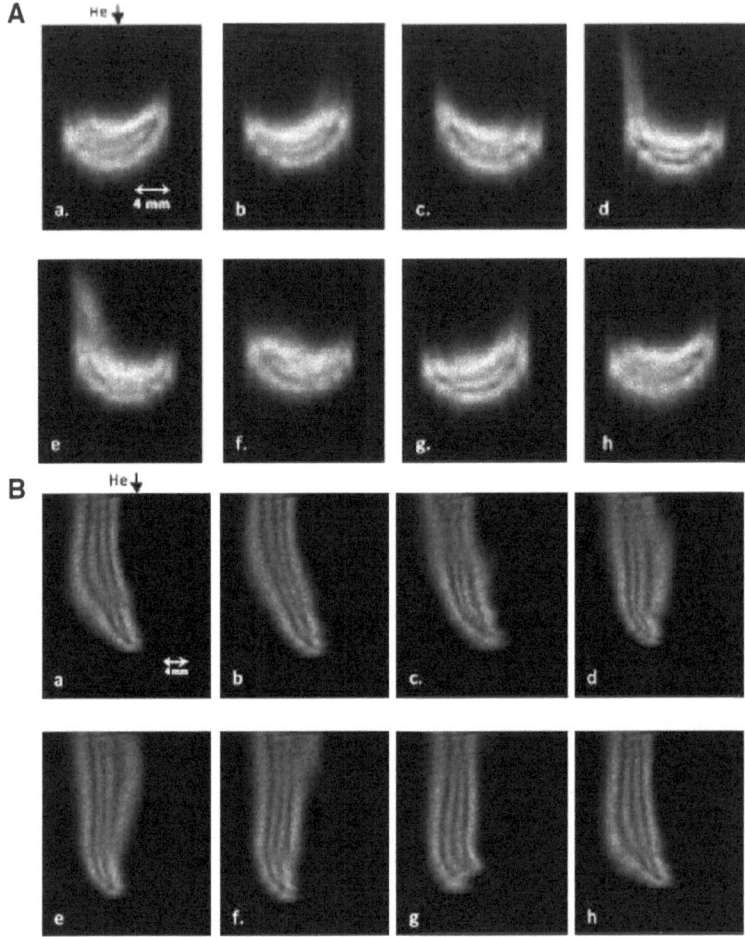

Fig. 5.7 (**A**) Cap (horizontal knife-edge) and (**B**) edge visualization (vertical knife-edge) of an unsteady jet issued through a 6-mm round nozzle at $Re = 53.28$ and $Fr = 4.83$. Schlieren images (**a–h**) are acquired at time intervals of 20 ms

4. *Sinuous oscillation*: This pattern is realized in the flapping mode when the ratio of the penetration length to jet diameter is large (Fig. 5.8i). It is the likely mode of unsteadiness for large Froude numbers.

In round nozzles, bobbing is the initial unstable mode observed at $Re = 35$. For higher Froude numbers, buoyancy is strengthened, and the jet has an appearance of a cap (regime C in Fig. 5.3). Thus, bobbing continues as the mode of fluctuation with increasing Froude number. For the same reason, this mode of instability is more common in smaller nozzle diameters. With increasing Reynolds number, jet penetration is longer, regimes A and B appear, and the jet becomes unsteady owing

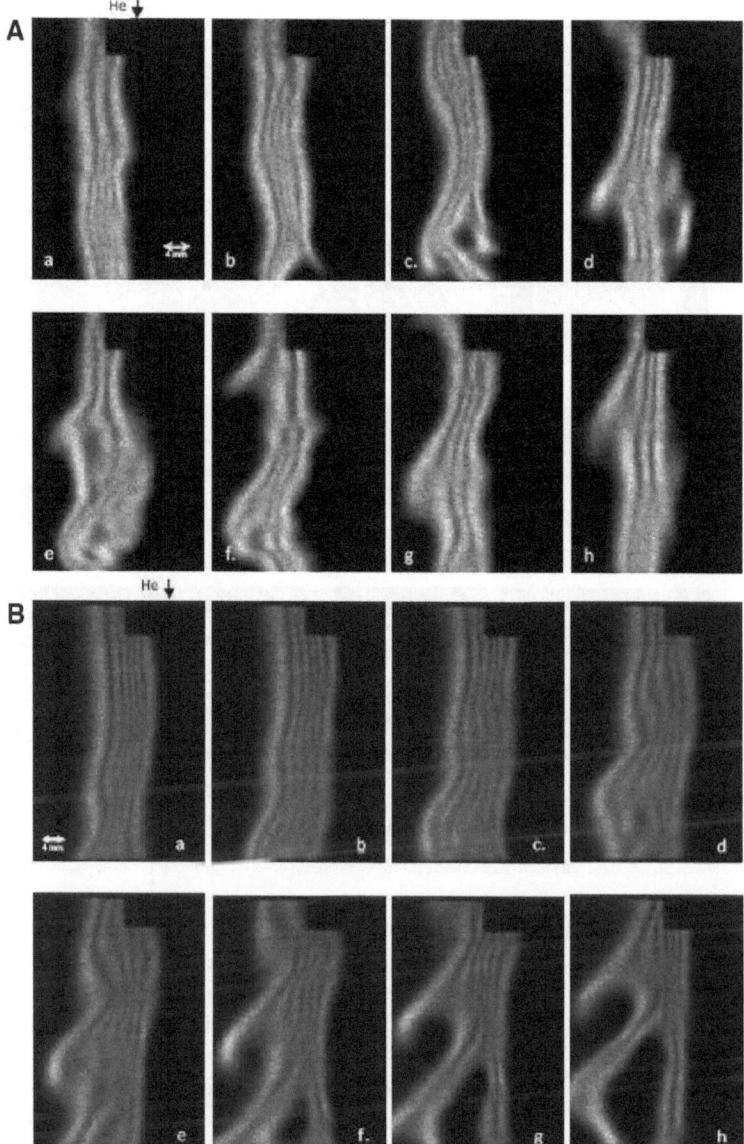

Fig. 5.8 Schlieren images with the knife-edge held vertical for a 6-mm round nozzle at (**A**) $Re =$ 76.11, $Fr = 6.89$ and a time interval of 20 ms; (**B**) $Re = 121.78$, $Fr = 11.03$ and a time interval of 60 ms

to Kelvin–Helmholtz instability—related to interfacial shear between the jet core and the upward moving annular flow. Spatially, these modes are visible in the annular region (regime B of Fig. 5.3) and is best visualized in schlieren by keeping

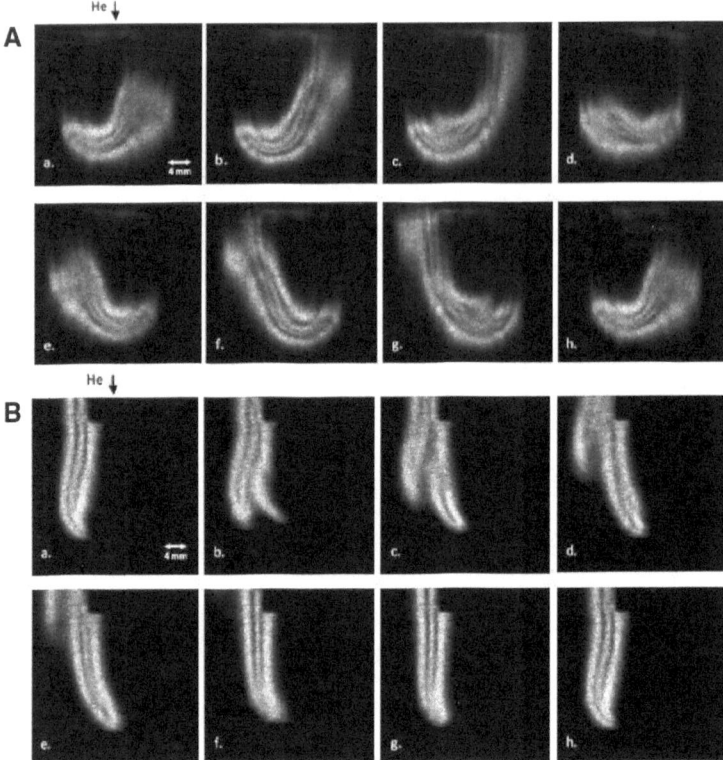

Fig. 5.9 (**A**) Cap and (**B**) edge visualization of a helium jet issued through a 6-mm square nozzle, $Re = 41.85$, $Fr = 3.79$. Images (**a–g**) are taken at time intervals of 16.66 ms, while image (**h**) is recorded after 0.167 s from image (**a**)

the knife-edge vertical. At intermediate values of the dimensionless parameters, multimode oscillations are possible. Gas diffusivity in air plays an important role because buoyancy forces depend on the gas concentration distribution in the flow field. Figures 5.7–5.10 also show that the jet may experience multimodal chaotic oscillations.

Passive control of jet mixing by adopting the correct nozzle geometry has been explored in the past for positively buoyant jets [28]. The literature on negatively buoyant jets is quite limited. Friedman et al. [7–9] reported no significant effect of nozzle geometry on the initiation of self-excited oscillations. Lin and Armfeld [14, 15] compared plane fountains and axisymmetric fountains numerically but did not comment on the transition to instability. The present study shows that a square nozzle intensifies fluctuations in a negatively buoyant jet. Although initiation of instability occurs at around the same Reynolds number as the round jet, the consequent magnitude of oscillations is greater for a square nozzle. Fluctuations start as pure flapping, while the change in penetration is small. The corresponding

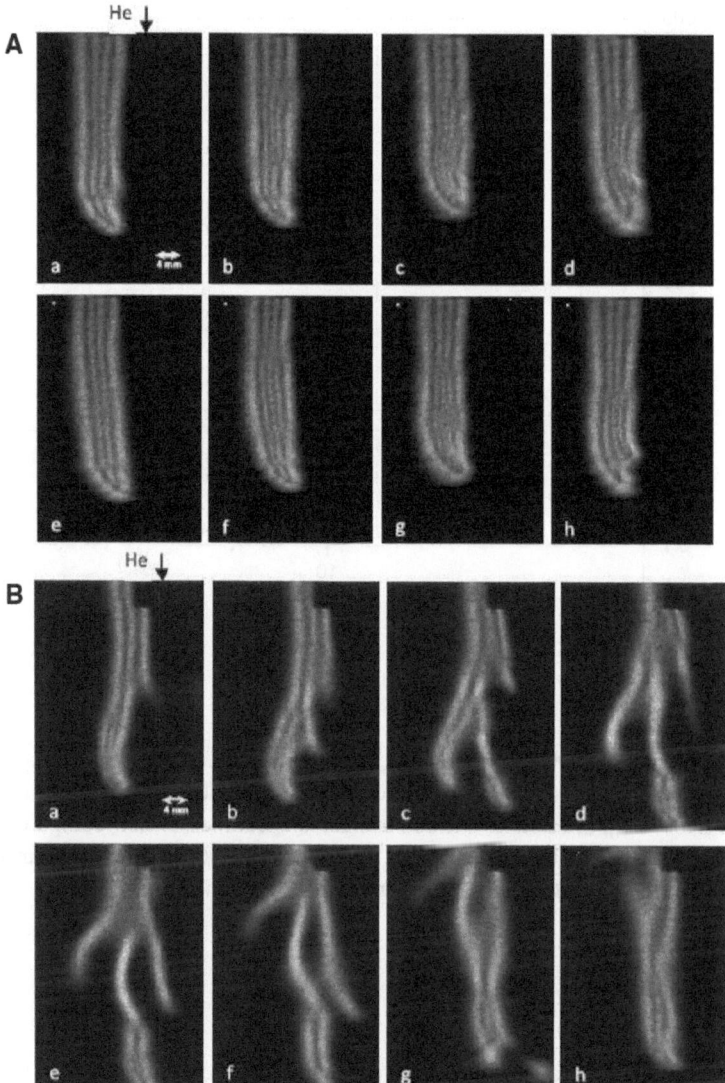

Fig. 5.10 Schlieren images (vertical knife-edge) for 6-mm round and square nozzles at comparable Froude and Reynolds numbers in the unsteady regime. (**A**) Round nozzle, $Fr = 60.89$, $Re = 5.51$, and (**B**) Square nozzle, $Fr = 5.41$, $Re = 59.78$. Fluctuations in the round jet are due to a combination of laminar bobbing and flapping, while for the square nozzle, the jet shows relatively strong fluctuations

oscillation for a round jet is in the form of gentle bobbing. On increasing the Froude number of a square jet, bobbing is also observed, but flapping remains the dominant mode.

Figure 5.9 shows schlieren images of the cap and the side region for $Re = 41.85$ and $Fr = 3.79$. Strong flapping motion is evident here. The near-stagnant fluid in the

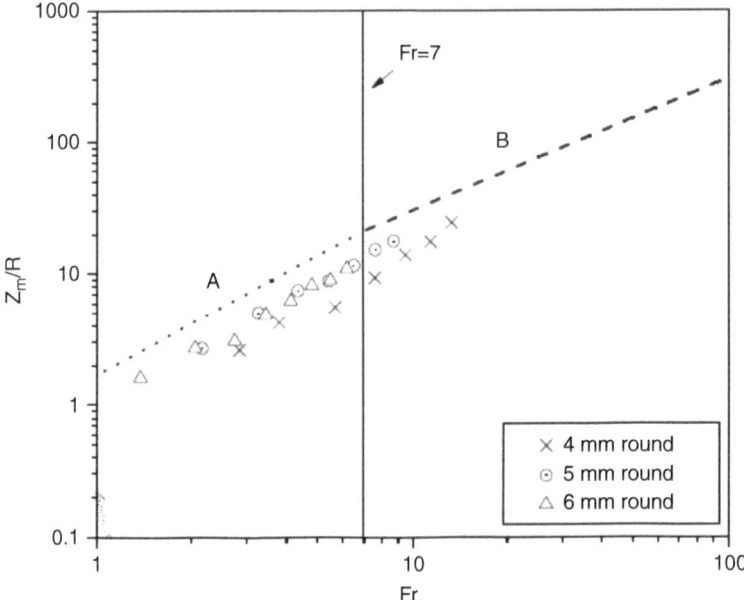

Fig. 5.11 Penetration length as a function of Froude number for various nozzle sizes compared with [35]. Portion (**A**): $Z_m/R = 1.7\, Fr^{1.3}$ and (**B**): $Z_m/R = 3.06\, Fr$

cap is periodically shed, and its size fluctuates with time. Figure 5.10 compares jets created by round and square nozzles at comparable Froude and Reynolds numbers. The jet from a square nozzle shows stronger unsteadiness when compared to the round. This result can be used to select square nozzles in place of circular nozzles when jet mixing with the ambient fluid needs to be maximized.

5.2.3 Penetration Distance

It is defined as the distance measured from the exit plane of the nozzle to the tip of the jet cap, along the central axis. Height and stability of laminar plane fountains are reported in [29]. Phipps et al. [26] measured penetration length on the basis of absolute helium concentration in a numerical simulation. This approach is not suitable in experiments since the helium diffusivity in air is quite large. Instead, the point of highest light intensity in a schlieren image has been adopted in the present study. Penetration length can properly be defined only for steady conditions of the jet. For an unsteady jet, the penetration length can be determined as an averaged quantity. The normalized penetration length of a jet Z_m scaled by the nozzle radius R is a function of the Froude and Reynolds numbers (Fig. 5.11).

Several studies use Froude number as the only relevant parameter for prediction of the penetration length. Others have used Richardson number, a mixed convection parameter in this respect. The data reported is in the turbulent regime where the effect of viscosity can be considered negligible compared to inertia forces. Figure 5.10 compares the penetration length recorded from the schlieren images of the present work with [35]. The Froude number range in the two studies is comparable. The Reynolds number range in the present work is quite small (≤ 200), while the range is 1,700–25,500 in the quoted reference. It is seen that the penetration length for the present data set is relatively smaller. In a later work, Zhang and Baddour [36] suggested that larger density difference between the jet fluid and ambient can reduce the penetration length. This work was restricted to the Boussinesq regime for which the density difference ratio was ≤ 0.1. In a helium jet, the density difference ratio is much higher, being ≈ 0.8. The strong interaction of a helium jet with the ambient fluid causes radial flow and a reduction in the penetration depth.

Previous literature has given importance to Froude number over Reynolds number in determining the penetration length. Friedman et al. [9] defined a corrected nondimensional number that combines Fr and Re in order to unify a top-hat velocity profile of turbulent flow with the parabolic profile of fully developed laminar flow. Philippe et al. [25] showed from experiments the dependence of penetration length individually on Re and Fr. Lin and Armfield [14] showed numerically that penetration depth is a function of these parameters, though different from the experiments of [25]. Figure 5.12 shows the plot of penetration distance as a function of the composite parameter $Fr \times Re^{0.5}$. Results show that the trend followed is

$$Fr \times Re^{0.5} = 0.23 \frac{Z_m}{R}.$$

This line is 33% below the prediction of [33]. Wardana et al. [32] published steady and unsteady images of a helium fountain formed by a 12-mm round nozzle in microgravity and normal gravity conditions. The penetration length information extracted from their images for normal gravity match the present data quite well.

Figure 5.12 compares the penetration length in round and square jets. The data for a square jet is seen to be slightly greater, compared to the round. Since the time-dependent oscillations for a jet from a square nozzle are stronger, this result is inconclusive and needs to be reviewed.

5.2.4 Power Spectra

Power spectra of light intensity signals are discussed in the present section. The presence of a dominant frequency as opposed to broad spectrum can be delineated from these studies. Spectra have been determined using FFT at distinct locations along the jet axis at a distance of R in the radial direction. Since the camera speed and aperture were adjusted to yield the best image sequence, the magnitudes of power spectrum cannot be directly compared. However, the onset of instability in the jet as a single harmonic and its progress toward a turbulent state can be examined.

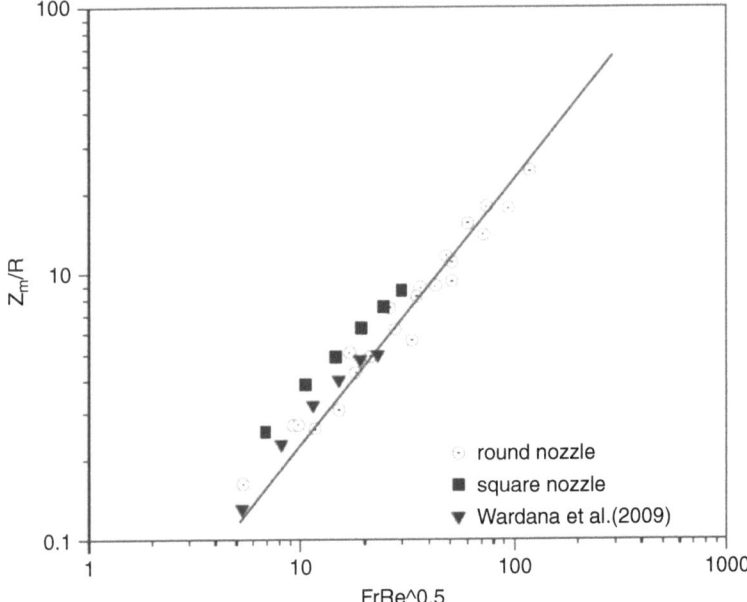

Fig. 5.12 Penetration length as a function of the composite parameter $Fr \times Re^{0.5}$. Penetration length is found to be slightly greater for a square nozzle. Comparison with [32] for a round jet shows a good match

Figure 5.13 compares power spectra recorded at various streamwise locations for two sets of Froude and Reynolds numbers. These are ($Fr = 4.13$, $Re = 45.67$) and ($Fr = 5.51$, $Re = 60.89$), while the nozzle diameter is 6 mm. The lower set of values (Fig. 5.13a) represents the threshold when unsteadiness sets in. Specifically, the initial spectra for $z/R \leq 5$ is broadband and noisy, while a sharp peak at 7 Hz (and a few of its harmonics) is seen at $z/R = 6$. This location corresponds to region C, the cap of the jet. The instability is not strong enough to affect the upstream region. Figure 5.13b shows power spectra for higher values of Froude and Reynolds numbers for $z/R = 0$–9. Here, spectral peaks as well as definite harmonics are visible over the entire length of the jet. The jet shows flapping instability, and spectral peaks are present along its entire length. The peak is the strongest in the cap region, namely, $z/R = 9$. Hence, the jet may be undergoing both flapping as well as bobbing motions.

It is to be expected that fluctuations would intensify at large Froude and Reynolds numbers. In this context, Fig. 5.14 shows power spectra for $Re = 102.75$ and $Fr = 17.09$. The nozzle diameter selected here is 4 mm since it allows a wider range of streamwise distances to be examined. The range covered in this experiment is $0 < z/R < 16$. The spectra in the near field are seen to be distinctly different from the far field. The near-field instability is inertia dominated and related to regime B where flow is annular and disturbances grow by the Kelvin–Helmholtz instability

Fig. 5.13 Power spectra of
light intensity signals at
various streamwise locations
of a jet issued through a round
nozzle of 6-mm diameter:
(a) $Re = 45.67$, $Fr = 4.13$
(b) $Re = 60.89$, $Fr = 5.51$

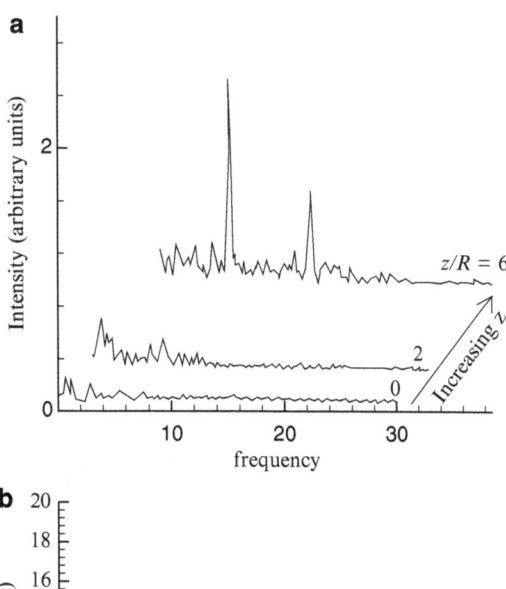

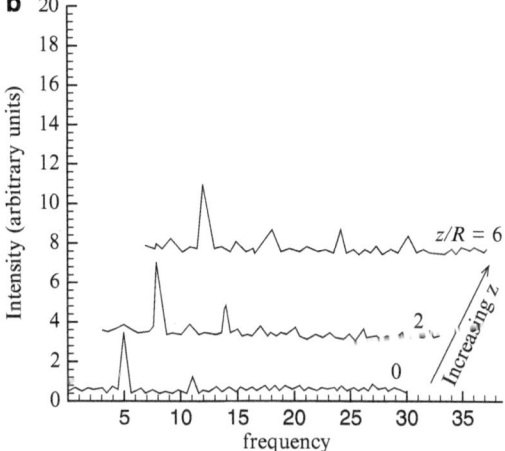

mechanism. The far field is dominated by gravity and unstable modes originate in the jet cap. Thus, in a general setting, the three regimes in the jet continue to make their presence felt in terms of the time-averaged field as well as fluctuations.

The unsteady behavior of a buoyant helium jet from a square nozzle is significantly different from a round nozzle. For a square nozzle, Fig. 5.15a shows the power spectra at various streamwise locations for $Fr = 3.35$ and $Re = 35.87$. Though the Froude number is small, fluctuations are seen over the entire length of the jet. This observation is supported by the flapping motion seen in the schlieren images of Fig. 5.9. For this set of parameters, region A is negligible. The radial outward velocity component causes flow to oscillate in the sideways direction, facilitating a common frequency throughout the jet field. The highest spectral peak is seen to occur in region B. For higher flow parameters, $Fr = 5.41$, $Re = 59.78$, Fig. 5.15b, a common dominant frequency is once again obtained at all streamwise locations. The

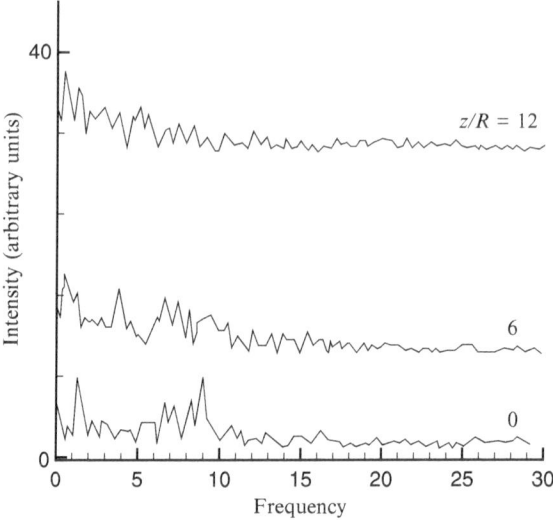

Fig. 5.14 Power spectra of light intensity at various streamwise locations (z/R) of a jet issued through a round nozzle of 4-mm diameter at $Re = 102.75$, $Fr = 17.09$

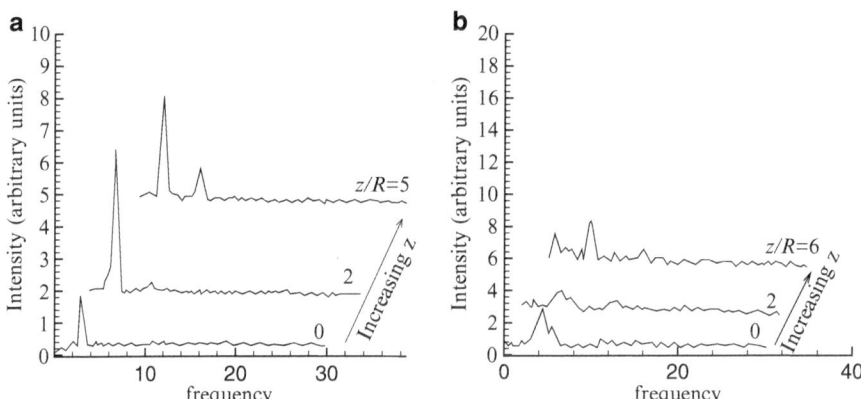

Fig. 5.15 Power spectra of light intensity at various streamwise locations (z/R) of a jet issued through a square nozzle of 6-mm hydraulic diameter: (**a**) $Re = 35.87$ and $Fr = 3.25$, (**b**) $Re = 59.78$, $Fr = 5.41$

spectral peak, however, shifts to region C. This is because flapping motion alone is expected for the lower range of parameters, while flapping and bobbing motions are jointly present at the higher end.

5.2.5   Strouhal Number

The power spectra of the previous section show the jet to contain dominant peaks at select frequencies. These frequencies can be cast in the form of a Strouhal number

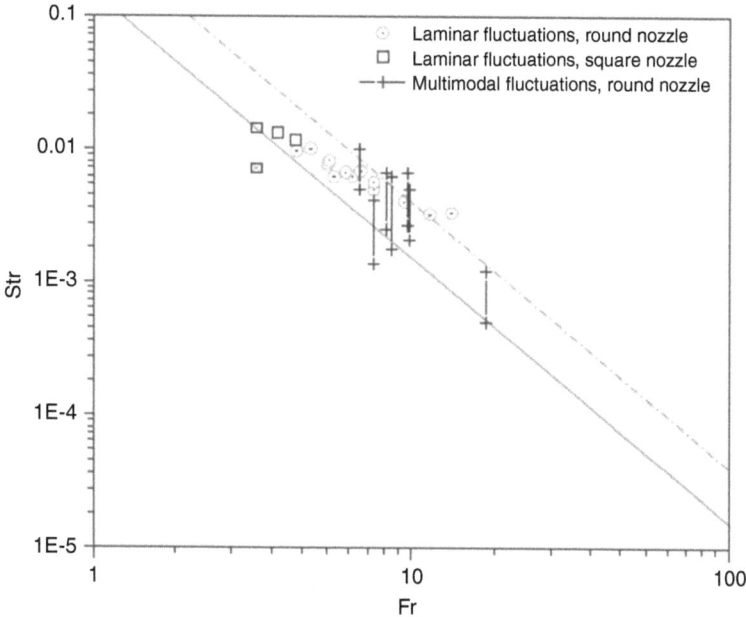

Fig. 5.16 Strouhal number calculated from the most dominant mode as a function of Froude number. *Straight lines* indicate previously reported bounds on Strouhal number, [33]

(Str) that in turn must scale with Froude number. Yildirim and Agrawal [34] reported self excited oscillation of momentum dominated helium jets. For buoyant plumes, Middleton [19] noted that frequency can be scaled as

$$\text{Str} = \frac{f \times R}{U} = \frac{g' \; R}{U \; U} \approx \frac{1}{(Fr)^2}. \tag{5.2}$$

The equivalent nondimensional time scale is given as

$$\tau = \frac{t \times g'}{U}.$$

Previous studies, [23], have also used these definitions of timescale and Strouhal number. Friedman et al. [6] described height oscillations by a general formula:

$$\text{Str} = \frac{fd}{U} = 0.16.$$

Figure 5.16 shows the plot of Strouhal number as a function of Froude number for the circular and square nozzle geometries. The data is compared with Williamson et al. [33] who gave bounds on Strouhal number as

$$\text{Str} = \frac{0.15}{Fr^2} \quad \text{(solid line)}$$

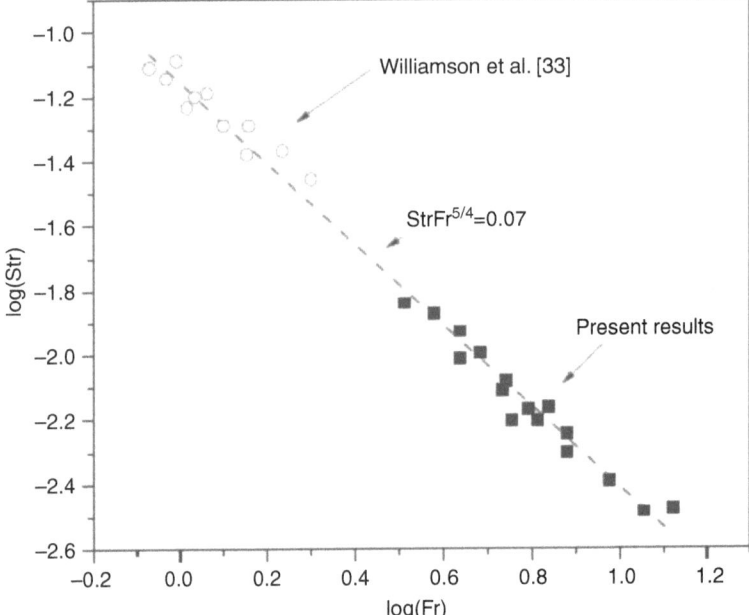

Fig. 5.17 Comparison of the empirical correlation Str $\times$ $Fr^{1.25}$ = 0.07 of the present study with data extracted from [33] for a laminar unsteady jet

and

$$\text{Str} = \frac{0.4}{Fr^2} \quad \text{(dashed-dotted line)}$$

corresponding, respectively, to the laminar and transitional regimes. The figure shows that the bounds hold for a non-Boussinesq fluid such as helium. Even multimodal behavior of the jet falls within the bounds prescribed.

Strouhal numbers of the present work for a laminar unsteady jet fall within the band referred above. A closer examination shows the data to follow a line with a distinct slope. This trend can be seen in the data of [33] as well, except that the data points are few in number. Such points were extracted and compared with the best fit of the present study. The best fit is given by the relation

$$\text{Str} \times Fr^{1.25} = 0.07$$

with a standard error of 0.9%. Figure 5.17 shows good agreement between the reported data and the correlation.

For a jet from a square nozzle, the dependence of Strouhal number on Froude number is similar to that of a round jet. Small differences are observed in Strouhal number at the onset of fluctuations. Since fluctuations in a square jet has flapping as well as bobbing components, it is not always true that the spectral peak matches the fundamental frequency. Occasionally, the highest peak could occur at one subharmonics. Thus, there could be departure from the linear trend of a round jet if the subharmonic frequency is considered as the dominant mode. This data is shown using crossed squares in Fig. 5.16. Nevertheless, for larger Froude numbers, the spectral peak coincides with the fundamental, and the data is shown as uncrossed squares in Fig. 5.16.

5.3 Multiple Jets

This section presents the interaction characteristics of two gas jets mixing under ambient conditions. The gases considered are helium that is lighter than air and oxygen, that is marginally heavier. The closely spaced nozzles (5.5-mm diameter) are sufficiently long (340 mm) so that the flows emerging from them are fully developed. Flow from the nozzles is oriented in the vertically downward direction. Both circular and square nozzles have been studied. The test section of the apparatus is octagonal and made of Plexiglas, similar to the one in Fig. 5.1. Glass windows are used for the sides that permit the passage of the light beam. The test cell is 340 mm long with a distance of 85 mm between parallel faces of the test cell. Flow rates of gases are measured using suitable rotameters. The range of parameters realized with the helium jet are summarized in Table 5.2. The parameters considered for oxygen are provided in Table 5.3. In the color schlieren images shown, suffix "1" is for oxygen and "2" is helium.

Helium being lighter than air, it is to be expected that a helium jet would form a cap, turn around and rise in the vertical direction (Sect. 5.2). Being heavier than air, oxygen would proceed downward like a free jet. With mixing, it is possible

Table 5.2 Experimental conditions for a helium jet

Nozzle shape	Nozzle diameter	Flow rate (in lpm)	Froude number	Reynolds number
Circular	5.5 mm	1	3.25	31.6
		1.5	4.88	47.4
		2	6.51	63.2
		3	9.76	94.8
		4	13.02	126.5
		5	16.27	158.9
		6	19.53	190.1
		7	22.78	221.3
Square	4.9	1	3.25	31.6
		1.5	4.88	47.4
		2	6.51	63.2
		3	9.76	94.8
		4	13.02	126.5
		5	16.27	158.9
		7	22.78	221.3

Table 5.3 Experimental conditions used for an oxygen jet	Nozzle shape	Nozzle diameter	Flow rate (in lpm)	Froude number	Reynolds number
	Circular	5.5 mm	1	36.64	248.4
			1.5	54.96	372.5
	Square	4.9	1	36.64	248.4
			1.5	54.96	372.5

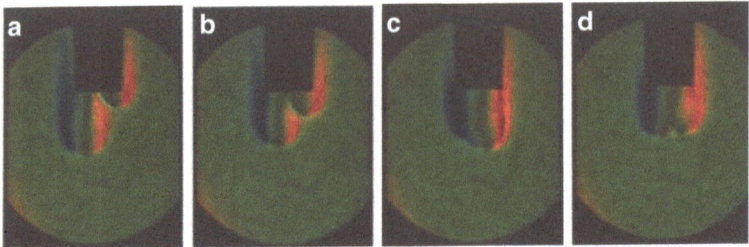

Fig. 5.18 Circular nozzle: color schlieren images where the oxygen jet parameters are $Re_1 = 79$, $Fr_1 = 8.13$ and for the helium jet, (**a**) $Re_2 = 31$, $Fr_2 = 3.25$ (**b**) $Re_2 = 48$, $Fr_2 = 4.88$ (**c**) $Re_2 = 63$, $Fr_2 = 6.51$ (**d**) $Re_2 = 79$, $Fr_2 = 8.13$

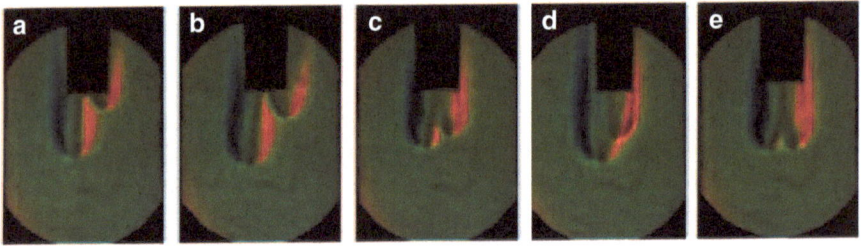

Fig. 5.19 Circular nozzle: color schlieren images where the oxygen jet parameters are $Re_1 = 95$, $Fr_1 = 9.76$ and for the helium jet, (**a**) $Re_2 = 31$, $Fr_2 = 3.25$ (**b**) $Re_2 = 48$, $Fr_2 = 4.88$ (**c**) $Re_2 = 63$, $Fr_2 = 6.51$ (**d**) $Re_2 = 79$, $Fr_2 = 8.13$ (**e**) $Re_2 = 95$, $Fr_2 = 9.76$

that the oxygen–helium mixture density is lower than air, and both jets rise upward. The depth of penetration of the two jets would depend on the respective Reynolds and Froude numbers of oxygen and helium, increasing with Reynolds number of oxygen. Under extreme conditions, helium can rise while oxygen continues to descend vertically downward. Thus, a variety of flow regimes are possible, some of which are explored in Figs. 5.18–5.21. Multiple jet interactions lead to complex three dimensionality, which can be better resolved using appropriate tomographic algorithms [27].

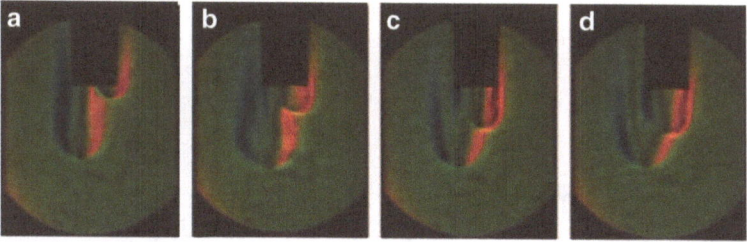

Fig. 5.20 Square nozzle: color schlieren images where the oxygen jet parameters are $Re_1 = 111$, $Fr_1 = 11.39$ and for the helium jet, (**a**) $Re_2 = 31$, $Fr_2 = 3.25$ (**b**) $Re_2 = 48$, $Fr_2 = 4.88$ (**c**) $Re_2 = 63$, $Fr_2 = 6.51$ (**d**) $Re_2 = 79$, $Fr_2 = 8.13$

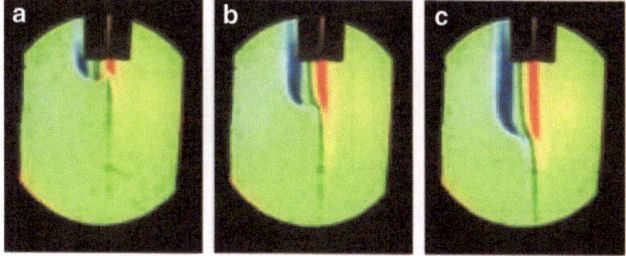

Fig. 5.21 Circular nozzle: Color schlieren images where the oxygen jet parameters are $Re_1 = 248$, $Fr_1 = 36.64$ and for the helium jet, (**a**) $Re_2 = 31$, $Fr_2 = 3.25$ (**b**) $Re_2 = 63$, $Fr_2 = 6.51$ (**c**) $Re_2 = 95$, $Fr_2 = 9.76$

5.4 Jet Impingement

Buoyant jets are encountered in many industrial applications. Buoyancy forces can arise either due to concentration difference or temperature difference between the jet fluid and the ambient. The presence of multiple jets leads to added complexity because of the instabilities triggered by the neighboring jets. These jets may impinge on a wall creating wall jets. The hydrodynamics of these flows and transport phenomena depend on the flow conditions and overall geometry of the nozzle with respect to the solid surfaces. Under real-life conditions, jets are unsteady with characteristic length and timescales in the near- and the far-field regions. Optical techniques are the most suited for characterizing flow structures and turbulent fluctuations in single and multiple jets. The wall temperature fluctuations and their correlation with jet instability and vortex structures are important in various contexts. The present section reports an application of the shadowgraphy technique for studying impingement of hot and cold water jets over a horizontal surface.

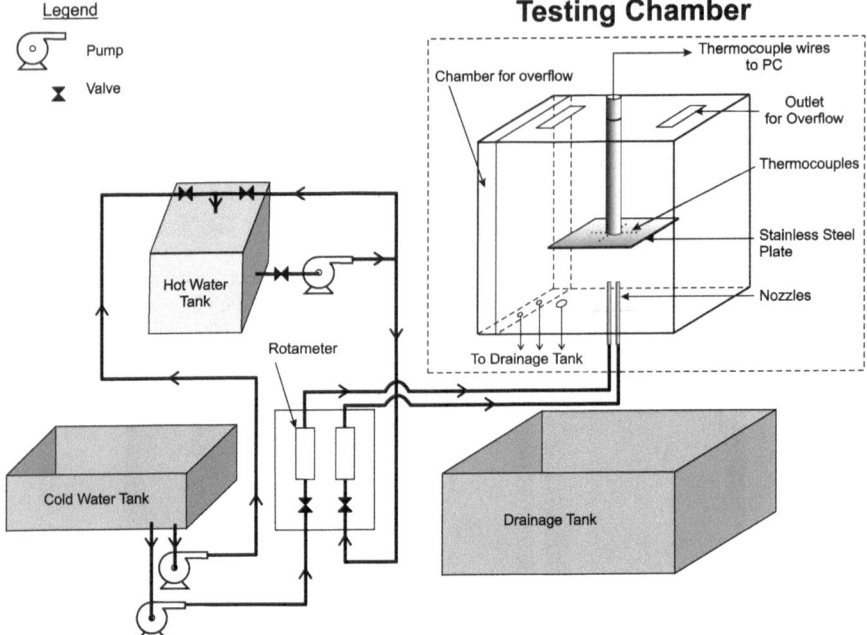

Fig. 5.22 Schematic diagram of the experimental setup. Hot and cold water jets enter the test chamber via rotameters and nozzles, and constant flow rates are maintained by individual pumps. The impingement plate is attached to a traversing mechanism that fixes its location with respect to the nozzle exit

5.4.1   Experimental Apparatus

The schematic diagram of the experimental setup is shown in Fig. 5.22. It consists of a hot water tank, a tank containing water at ambient temperature (cold water tank) and a control panel with three voltage controllers (15 amp) and two rotameters. The hot water tank is thermostatically controlled to within ± 1 K. The hot and the cold streams emerging from nozzles of equal diameter but adjustable flow rates mix in the test chamber, before impinging on a horizontal surface kept above. Distilled, demineralized water is used in all the experiments. Typical flow rates are 2–6 lpm, and temperature differences are of the order of 20 K. The corresponding Reynolds number is around 24,000, while the Richardson number is around 0.05, indicating a weak mixed convection flow.

The testing chamber houses two nozzles of 9-mm diameter. This chamber is made of transparent Plexiglas and permits imaging and visualization of the flow field. The impingement plate, made of stainless steel, can be moved toward or away from the nozzles by a traversing mechanism. The impingement plate houses 51 copper–constantan thermocouples (type T) and is connected to a computer via a temperature measurement and data acquisition card (*National Instruments*).

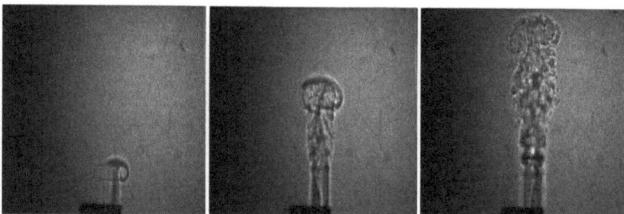

Fig. 5.23 Flow structures of a heated water jet in an environment of water in the initial stages before the jet hits the impingement plate. The time separation of images shown is 1.5 s

Thermocouples are placed in the plate through 1-mm-diameter holes in the plate. Thermocouple beads are just exposed to the flow so as to pick up temperature fluctuations. The correlation between spectra of wall temperature fluctuations and those of the jet is a subject of considerable importance.

5.4.2 Flow Distribution and Light Intensity Spectra

The mixing of cold and hot water jets is visualized using a shadowgraph method. The He-Ne laser is used as a light source. The laser output, suitably expanded to a collimated beam, passes through the test chamber, and the intensity-modulated output is recorded using a high-speed monochrome CCD camera (*MCI 1302*). Since density gradients are heightened by schlieren and shadowgraph techniques, structures appearing in the images discussed below can be interpreted as vortices, to a first approximation. For the parameters considered in the present section, length and timescales could be resolved at a camera speed of 250 frames per second. In the time domain, light intensity fluctuations develop due to those in velocity and hence temperature. These may be damped when the overall fluid temperature diminishes, for example, when the cold jet flow rate is increased. In the images presented, the cold jet is at the surrounding fluid temperature and is not visible. The heated jet and its mixing with the surrounding medium is clearly revealed. For comparison with starting gas jets discussed in Sect. 5.2.2.1, the initial evolution of heated water jets are shown in Fig. 5.23.

Figure 5.24 presents instantaneous visualization images when cold flow is zero and only the hot jet at flow rates of 2 and 6 lpm prevail. These images represent a sample extracted from a long video sequence that includes initial jet instabilities to the formation of wall jets and finally a large-scale circulation pattern in the experimental chamber. In this chapter, the temperature difference between the hot jet and the ambient fluid is 20 K. The width of the jet increases in the axial direction. Hence, for increasing separation between the nozzle exit and the impingement plate, a weaker jet in terms of momentum flux strikes the impingement plate. The jet is bound to be cooler as well. It is to be seen that the interface deformation between jet

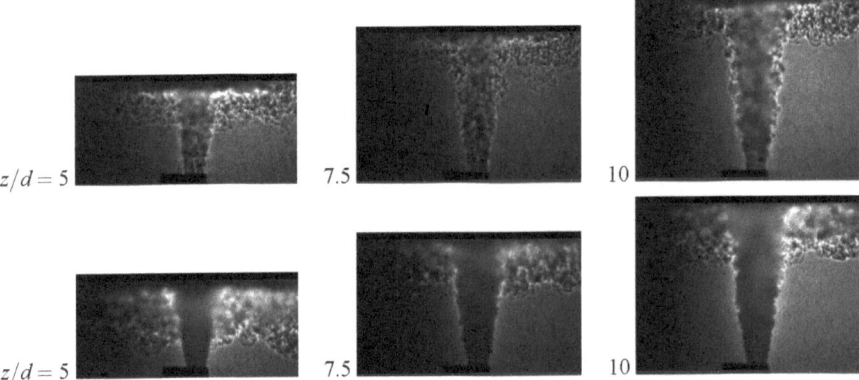

Fig. 5.24 Instantaneous shadowgraph images for various flow rates of the hot water jet and impingement plate locations (z/d). First row (*top*): $Q_c = 0$ lpm and $Q_h = 2$ lpm; second row: $Q_c = 0$ lpm and $Q_h = 6$ lpm

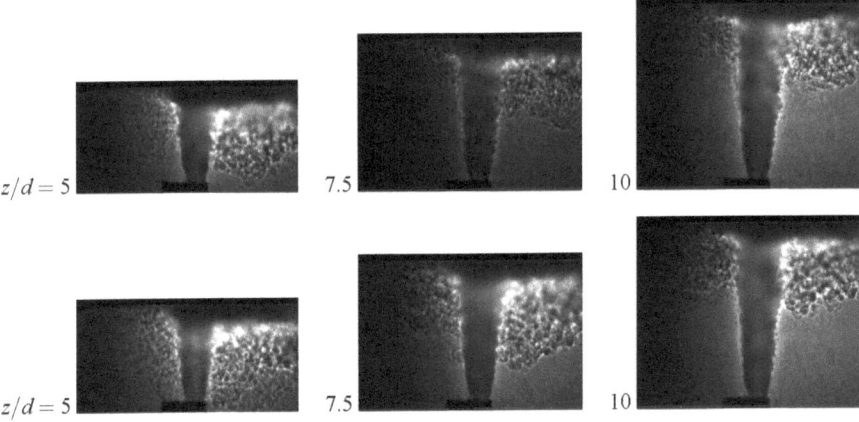

Fig. 5.25 Instantaneous shadowgraph images for equal hot and cold jet flow rates as a function of the location (z/d) of the impingement plate. First row (*top*): $Q_c = 4$ lpm and $Q_h = 4$ lpm; second row: $Q_c = 6$ lpm and $Q_h = 6$ lpm

and the ambient is higher for low flow rates. It is an indication of greater role played by the buoyancy compared to inertia for initiation of interfacial instability. Increase in the velocity of the hot jet shows greater influence on the spread of the wall jet.

Figure 5.25 shows shadowgraph images for equal flow rates of cold and hot water. The separation distance between the nozzle exit (below) and the impingement plate (above) is varied. Symmetric flow structures observed in Fig. 5.24 are no more visible after introducing the cold flow. The wall jet on the impingement plate also shows different spreading behavior on each side of the nozzle centerline. The thermal boundary layer thickness of the wall jet is higher on the hot side. The

Fig. 5.26 Shadowgraph images for an impingement plate location of $z/d = 10$ with unequal hot and cold jet flow rates. (**a**) $Q_c = 0\,\text{lpm}$, $Q_h = 6\,\text{lpm}$; (**b**) $Q_c = 2\,\text{lpm}$, $Q_h = 6\,\text{lpm}$; (**c**) $Q_c = 6\,\text{lpm}$, $Q_h = 6\,\text{lpm}$

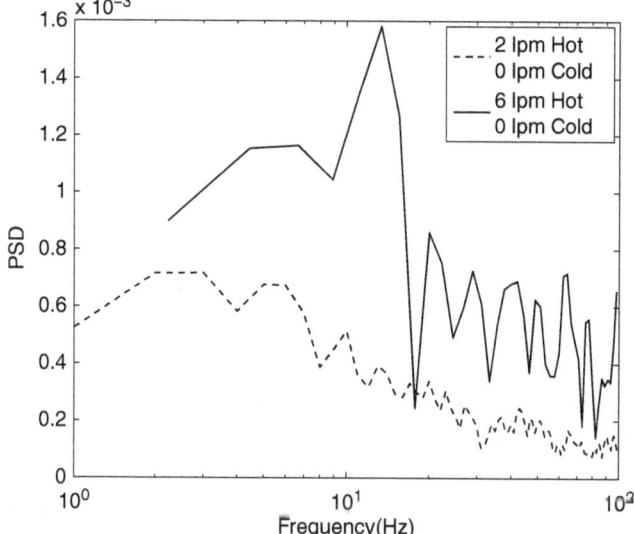

Fig. 5.27 Power spectral density (PSD) of light intensity with cold flow set to zero. Various flow rates of the hot jet are considered. The impingement plate is located at $z/d = 10$ from the exit plane of the nozzle

spread of thermal energy by the impingement plate can cause large-scale circulation in the main tank of the apparatus.

Figure 5.26 shows visualization images of impingement for two different velocity ratios and a given position of the impingement plate. The structure of the wall jet is clearly seen to be a function of the velocity ratio. The mixing between the hot and cold jet in the impingement region leads to the generation of flow structures with new length scales. These can be identified and measured by using appropriate image analysis techniques.

Figure 5.27 shows power spectra of light intensity fluctuations as a function of the flow rate of the hot jet. These have been determined at the interface region between the hot jet and the ambient fluid, the axial location being $z/d = 5$. The figure shows a dominant peak at about 10 Hz when the flow rate is increased from 2 to 6 lpm. The energy content also increases with flow rate, indicating an overall increase in the fluctuation intensity.

5.5 Summary

The studies reported in this chapter show that schlieren images can be used for qualitative and quantitative characterization of buoyant helium jets. The measurements with a helium jet match those of a salt-water fountain under similar flow conditions. Similarly, color schlieren images of multiple gas jets and shadowgraph images of hot and cold water jets impinging on a flat surface reveal interesting flow features.

References

1. W.D. Baines, J.S. Turner and I.H. Campbell, Turbulent fountains in an open chamber, Journal of Fluid Mechanics, Vol. 212, pp. 557–592, 1990.
2. W.D. Baines, A.F. Corriveau and T.J. Reedman, Turbulent fountains in a closed chamber, Journal of Fluid Mechanics, Vol. 255, pp. 621–646, 1993.
3. L.J. Bloomfield and R.J. Kerr R.J., A theoretical model of a turbulent fountain, Journal of Fluid Mechanics, Vol. , pp., 2000.
4. P.F. Crapper and W.D. Baines, Some remarks on non-Boussinesq forced plumes, Atmospheric Environment, Vol. 11. pp. 415–420, 1977.
5. R.W. Cresswell and R.T. Szczepura, Experimental investigation into a turbulent jet with negative buoyancy, Physics of Fluids, Vol. 5, pp. 2865–2878, 1993.
6. P.D. Friedman, Oscillation height of a negatively buoyant jet, J. Fluids Eng. Trans. ASME, Vol. 128, pp. 880–882, 2006.
7. P.D. Friedman and J. Katz, The flow and mixing mechanisms caused by the impingement of an immiscible interface with a vertical jet, Physics of Fluids, Vol. 11, pp. 2598–2606, 1999.
8. P.D. Friedman and J. Katz, Rise height for negatively buoyant fountains and depth of penetration for negatively buoyant jets impinging on an interface, J. Fluids Eng. Trans. ASME, Vol. 122, pp. 779–782, 2000.
9. P.D. Friedman, V. Vadakoot, W.J. Meyer and S. Carey, Instability threshold of a negatively buoyant fountain, Experiments in Fluidis, Vol. 42, pp. 751–759, 2007.
10. N.A. Jaffe, The laminar boundary layer with uniform injection of a foreign gas, Proc. Royal Soc. London, Vol. 317, pp. 393–405, 1970.
11. A. Jiaojain, L. Wing-Keung and S.C.M. Yu, On Boussinesq and non-Boussinesq starting forced plumes, Journal of Fluid Mechanics, Vol. 558, pp. 357–386, 2006.
12. E. Kaminski, S. Tait, and G. Carazzo, Turbulent entrainment in jets with arbitrary buoyancy, Journal of Fluid Mechanics, Vol. 526, p.361–376, 2005.
13. D.M. Kyle and K.R. Sreenivasan, The instability and breakdown of a round variable-density jet, Journal of Fluid Mechanics, 1993.
14. W. Lin and S.W. Armfield, The Reynolds and Prandtl number dependence of weak fountains, Journal of Computational mechanics, Vol. 31, pp. 379–389, 2003.
15. W. Lin and S.W. Armfield, Weak fountains, Journal of Fluid Mechanics, Vol. 403, pp. 67–88, 2000.

16. E.J. List, Turbulent Jets and Plumes, Annual Review of Fluid Mechanics, Vol.14, pp. 189–212, 1982.

17. T.J. McDougall, Negatively buoyant vertcial jets, Tellus, Vol. 33(3), pp. 313–320, 1981.

18. G. Michaux and O. Vauquelin, Solutions for turbulent buoyant plumes rising from circular sources, Physics of Fluids, Vol. 20, 066601, 2008.

19. J.H. Middleton, Times of rise for turbulent forced plumes, Tellus, Vol.31, pp. 82–88, 1979.

20. T. Mizushina, F. Ogino, H. Takeuchi and H. Ikawa, An experimental study of vertical turbulent jet with negative buoyancy, Warme-und Stoffubertragung, Vol. 16, p. 15, 1982.

21. D.E. Mowbray, The use of schlieren and shadowgraph techniques in the study of flow patterns in density stratified liquids, Journal of Fluid Mechanics, Vol. 27(3), pp. 595–608, 1967.

22. J. W. Nichols, P. J. Schmid and J. J. Riley, Self-sustained oscillations in variable-density round jets, Journal of Fluid Mechanics , Vol. 582, pp. 341–376, 2007.

23. L. Pantzlaff and R.M. Lueptow, Transient positively and negatively buoyant turbulent round jets, Expt. Fluids, Vol. 27, pp. 117–125, 1999.

24. P.N. Papanicolaou, I.G. Papakonstantis and G.C. Christodoulou, J. Fluid Mech, Vol. 614, pp. 447–470, 2008.

25. P. Philippe,C. Raufaste, P. Kurowski and P. Petitjeans, Penetration of a negatively buoyant jet in a miscible liquid, Phys. Fluids., Vol. 17, pp. 1–10, 2005.

26. M.R. Phipps, Y. Jaluria and T. Eklund, Helium-based simulation of smoke spread due to fire in enclosed spaces, Journal of Combustion Science and Technology, Vol. 157, pp. 63–86, 1997.

27. A.T. Ramsey and M. Diesso, Abel inversions: error propagation and inversion reliability, Review of Scientific Instruments, Vol. 70, pp. 380–383, 1999.

28. M. Salewski, D. Stankovic and L. Fuchs, Mixing in circular and non-circular jets in cross-flow, Flow, Turbulence and Combustion, Vol. 80(2), pp. 255–283, 2008.

29. N. Srinarayana, G.D. McBain, S.W. Armfield and W.X. Lin, Height and stability of laminar plane fountains in a homogeneous fluid, International Journal of Heat and Mass Transfer, Vol. 50 (7–8). pp. 1592–1602, 2008.

30. K.R. Sreenivasan, S. Raghu and D. Kyle, Absolute instability in variable density round jets, Experiments in Fluids, Vol. 7, pp. 309–317, 1989.

31. J.S. Turner, Jets and plumes with negative or reversing buoyancy, J. Fluid Mech., Vol. 26, pp. 779–792, 1966.

32. I.N.G. Wardana, H. Kawasaki and T. Ueda, Near-field instability of variable property jet in normal gravity and microgravity fields, Experiments in Fluids, Vol. 47, pp. 239–249, 2009.

33 N. Williamson, N. Srinarayana, S.W. Armfield, G.D. Mcbain and W. Lin, Low-Reynolds-number fountain behaviour, Journal of Fluid Mechanics, Vol. 608, pp. 297–317, 2008.

34. B.S. Yildirim and A.K. Agrawal, Full-field measurements of self-excited oscillations in momentum-dominated helium jets, Experiments in Fluids, Vol. 38, pp. 161–173, 2005.

35. H. Zhang and R.E. Baddour, Maximum penetration of vertical round dense jets at small and large Froude numbers, J. Hydraulic Eng., Vol. 124, pp. 550–553, 1998.

36. H. Zhang and R.E. Baddour, Density effect on round turbulent hypersaline fountain, Journal of Hydraulic Engineering, Vol. 135(1), pp. 57–59, 2009.

Chapter 6
Closure

6.1 Introduction

This monograph describes optical imaging of thermal and solutal concentration fields in several interesting applications. Depending on the analytical tools used, optical images can yield a considerable amount of information. It is to be expected that these imaging techniques would find greater applicability in the characterizing complex phenomena. Future directions in refractive index-based imaging and a collection of recent references [1–17] are presented in the following sections.

6.2 Future Directions

Refractive index-based measurements are possible only in a transparent medium; fluid regions with particulates cannot be imaged by this route. Refractive index techniques, however, perform whole-field measurements, a significant advantage over methods based on scattering. In fact, one can see the evolution of these methods towards making time-dependent three-dimensional measurements a reality.

Future developments in refractive index methods are likely to see the following trends:

1. Utility in field-scale applications, satellite-level imaging of natural phenomena, for example.
2. Imaging of transport phenomena at microscales, in MEMS, for example.
3. Combined measurements using, say, schlieren (in high gradient regions) and interferometry (in regions of low gradients).
4. Usage of multiple wavelengths of the laser including white light sources.
5. Extension to measurement with other radiation sources.
6. Superior analytical tools for image analysis.
7. Surface topography measurements.

P.K. Panigrahi and K. Muralidhar, *Imaging Heat and Mass Transfer Processes*, 131
SpringerBriefs in Applied Sciences and Technology 4, DOI 10.1007/978-1-4614-4791-7_6,
© Pradipta Kumar Panigrahi and Krishnamurthy Muralidhar 2013

8. Motion camera schlieren measurements of moving objects such as rockets.
9. Phase-shifting schlieren method for situations where refractive index variations are large.

Schlieren, in particular, will see developments in velocity measurement where light intensity is used as a tracer. The absence of inertia and the availability of high-speed cameras will make turbulence measurements simpler. Bulk property measurements such as thermal diffusivity in complex fluids and biological solutions is an emerging area of research.

References

1. D. Beghuin, J-L. Dewandel, L. Joannes, E. Foumouo, and P. Antoine, Optics Letters, Vol. 35(22), pp. 3745–3747, 2010.
2. F. Croccolo and D. Brogioli, Quantitative Fourier analysis of schlieren masks: the transition from shadowgraph to schlieren, Applied Optics, Vol. 50(20), pp. 3419–3427, 2011.
3. M.J. Hargather and G.S. Settles, Natural background-oriented schlieren imaging, Experiments in Fluids, Vol. 48, pp. 59–68, 2010.
4. J. G. Fujimoto and D. Farkas, editors, *Biomedical Optical Imaging*, Oxford University Press, Oxford, UK, 2009.
5. M.J. Hargather and G.S. Settles, A comparison of three quantitative schlieren techniques, Special issue of Optics and Lasers in Engg., Vol. 50(1), pp. 8–17, 2012.
6. E.D. Iffa, A.R.A. Aziz and A.S. Malik, Velocity field measurement of a round jet using quantitative schlieren, Applied Optics, Vol. 50(5), pp. 618–625, 2011.
7. T. Kirmse, J. Agocs, A. Schroeder, J. Martinez Schramm, S. Karl, and K. Hannemann, Application of particle image velocimetry and the background oriented schlieren technique in the high enthalpy shock tunnel Goettingen, Shock Waves, Vol. 21, pp. 233–241, 2011.
8. F. Moisy, M. Rabaud, and K. Salsac, A synthetic schlieren method for the measurement of topography of a liquid interface, Experiments in Fluids, Vol. 46(6), pp. 1021–1036, 2009.
9. Kai-Erik Peiponen, R. Myllyla and A. V. Priezzhev, *Optical Measurement Techniques: Innovations for Industry and the Life Sciences*, Springer series in Optical sciences, Ed. W. T. Rhodes, Springer, Berlin, 2009.
10. Jan-Pierre Prenel and Dario Ambrosini, Editors, *Advances in Flow Visualization,* Special issue of Optics and Lasers in Engg., Vol. 50(1), pp. 1–98, 2012.
11. S. Prasad, L. M. Bruce and J. Chanussot, editors, *Optical Remote Sensing: Advances in Signal Processing and Exploitation Techniques*, series on Augmented vision and reality, Ed. R. Hammoud, Springer, Berlin, 2011.
12. M. Raffel, R. Hernandez-Rivera, B. Heine, A. Schroeder, and K. Mulleners, Density tagging velocimetry, Experiments in Fluids, Vol. 51, pp. 573–578, 2011.
13. R. P. Satti, P.S. Kolhe, S. Olcmen, and A.K. Agrawal, Miniature rainbow schlieren deflectometry system for quantitative measurements in micro-jets and flames, Applied Optics, Vol. 46(15), pp. 2954–2962, 2007.
14. R. S. Sirohi, *Optical Methods of Measurement: Wholefield Techniques*, series on Optical science and engineering, Ed. B. J. Thompson, CRC Press, Boca Raton, 2009.
15. A. Srivastava, K. Tsukamoto, E. Yokoyama, K. Murayama and M. Fukuyama, Fourier analysis based phase shift interferometric tomography for three-dimensional reconstruction of concentration field around a growing crystal, Journal of Crystal Growth, Vol. 312, pp. 2254–2262, 2010.

16. L.M. Weinstein, Review and update of lens and grid schlieren and motion camera schlieren, Eur. Phys. J. (special topics), Vol. 182, pp. 65–95, 2010.
17. B. Zhang, Y. He, Y. Song, and A. He, Deflection tomographic reconstruction of a complex flow field from incomplete projection data, Optics and Lasers in Engineering, Vol. 47, pp. 1183–1188, 2009.